The Future of the Northern Forest

The Future of the Northern Forest

Edited by

Christopher McGrory Klyza
Stephen C. Trombulak

Middlebury College Press
Published by University Press of New England
Hanover and London

Middlebury College Press
Published by University Press of New England, Hanover, NH 03755

Printed in the United States of America
5 4 3 2 1

UNIVERSITY PRESS OF NEW ENGLAND
publishes books under its own imprint and is the publisher for Brandeis University Press, Brown University Press, University of Connecticut, Dartmouth College, Middlebury College Press, University of New Hampshire, University of Rhode Island, Tufts University, University of Vermont, Wesleyan University Press, and Salzburg Seminar.

Library of Congress Cataloging-in-Publication Data
The Future of the northern forest / edited by Christopher McGrory Klyza and Stephen C. Trombulak.
p. cm.
Includes bibliographical references (p.) and index.
ISBN 0-87451-682-X (cl) —ISBN 0-87451-710-9 (pa)
1. Forests and forestry—New England. 2. Forests and forestry—New York (State) 3. Forest policy—New England. 4. Forest policy—New York (State) 5. Forest ecology—New England. 6. Forest ecology—New York (State) 7. Forest products industry—New England. 8. Forest products industry—New York (State) I. Klyza, Christopher McGrory. II. Tombulak, Stephen C.
SD144.N37F88 1994
333.75'0974—dc20 94-17794
∞

This book was printed by Thomson-Shore, Inc. on recycled acid-free text stock, using soy-based inks.

Contents

Contributors

Emily M. Bateson is the Land Project Director of the Conservation Law Foundation. Address: Conservation Law Foundation, 62 Summer Street, Boston, MA 02110.

Dee Brightstar is an Abenaki artist, Tribal Council member, and a member of the Abenaki Research Project. Address: P.O. Box 316, Fairfax, VT 05454.

Thomas Carr is an Assistant Professor of Economics at Middlebury College, where he teaches and writes on environmental and natural resource economics. Address: Department of Economics, Middlebury College, Middlebury, VT 05753.

John Collins is Chair of the Adirondack Park Agency, a position to which he was appointed in September 1992. He has served on the APA board since 1984. He is a fifth generation Adirondack resident and was chair of the Town of Indian Lake Planning Board from 1965 through 1984. Address: Adirondack Park Agency, P.O. Box 99, Ray Brook, NY 12977.

John Elder is a Professor of English and Environmental Studies at Middlebury College. He is the author of *Imagining the Earth: Poetry and the Vision of Nature* and is a co-editor of *The Norton Book of Nature Writing*. He is currently at work on a book exploring the relationship between nature and culture in New England. Address: Department of English, Middlebury College, Middlebury, VT 05753.

Stephanie Kaza is Assistant Professor of Environmental Studies at the University of Vermont, where she teaches and writes on environmental ethics. She is the author of the recently published book *The Attentive Heart: Conversations with Trees*. Address: Environmental Program, The Bittersweet, University of Vermont, 153 South Prospect Street, Burlington, VT 05401.

Christopher McGrory Klyza is an Assistant Professor of Political Science at Middlebury College. He teaches environmental policy and has written on public lands issues. Address: Department of Political Science, Middlebury College, Middlebury, VT 05753.

John Moody is an ethnohistorian who has worked with the Abenaki Nation for the last fifteen years. Address: RFD, Sharon, VT 05065.

Tom Obomsawin is a singer and writer, as well as a member of the Abenaki Tribal Council. Address: Abenaki Nation of Missisquoi, P.O. Box 276, Swanton, VT 05488.

Carl Reidel is Director of the Environmental Program, Daniel Clarke Sanders Professor of Environmental Studies, and Professor of Natural Resources and Public Administration at the University of Vermont. He is a Past President of the American Forestry Association, and is currently the Vice Chair of the National Wildlife Federation Board of Directors, a Trustee for the New England Natural Resources Council, and Vice Chair of the Governor's Council of Environmental Advisors in Vermont. Address: Environmental Program, The Bittersweet, University of Vermont, 153 South Prospect Street, Burlington, VT 05401.

Hilda Robtoy is an Abenaki grandmother, Tribal Council member, and a member of the Abenaki Research Project. Address: 17 Greenwich Street, Swanton, VT 05488.

Jamie Sayen is the founder of Preserve Appalachian Wilderness and the editor of the *Northern Forest Forum*. Address: P.O. Box 52, Groveton, NH 03582.

Henry Swan is President of Wagner Woodlands, Inc., a company specializing in timberland investment, forest management, and consulting on forest-related ventures. His background combines training and experience in forestry, business, investment, and financial management. He is past Chairman of the Board of Trustees of the Society for the Protection of New Hampshire Forests and serves on their Lands Committee. He is also Director of a specialized forest products company. He was a member of the Governors' Task Force on Northern Forest Lands. He graduated from the University of Maine with a B.S. in Forestry in 1957 and from Harvard University with an M.B.A. in 1963. Address: Wagner Woodlands, Inc., P.O. Box 128, Lyme, NH 03768.

Stephen C. Trombulak is a Professor of Biology and Environmental Studies at Middlebury College, and is the former Director of the Environmental Studies Program. He is on the Board of Directors of the Vermont Natural Resources Council and is a contributing editor to the *Northern Forest Forum*. Address: Department of Biology, Middlebury College, Middlebury, VT 05753.

Brendan J. Whittaker is a professional forester, presently heading the Vermont Natural Resources Council's Northern Forest Project. In the past he has served as Vermont's Secretary of Environmental Conservation (1978 to 1985). Also an Episcopal clergyman since 1966, he presently is pastor of St. Mark Parish in Groveton, New Hampshire. He is currently Chair of the Selectboard of Brunswick, Vermont, where he has previously served as Town Lister, Zoning Administrator, and a member of the Planning Commission. He is also currently serving as one of the Vermont members of the Northern Forest Lands Council. Address: Vermont Natural Resources Council, 9 Bailey Ave., Montpelier, VT 05602.

Jonathan Wood is the forester for Bell-Gates Lumber Corporation in Jeffersonville, Vermont. He is a member of the Stewardship and Forest Legacy Committees for the State of Vermont, the Forest Policy Committee of the Green Mountain Division of the Society of American Foresters, the Multiple Use Association, and the Lamoille County Planning Commission, and is on the Board of Directors of the Vermont Forest Products Association. Address: Bell-Gates Lumber Corporation, P.O. Box 279, Jeffersonville, VT 05464.

Acknowledgments

We would like to thank a number of institutions and individuals for their help in bringing this book to completion. We thank our colleague John Elder who helped to organize a conference on the future of the Northern Forest held at Middlebury College in April 1992. It was this conference that served as the catalyst for undertaking this project, though by no means is the book a conference proceedings (only five chapters are based on conference presentations). Relatedly, we would like to thank Middlebury College, which sponsored the conference through the President's Fund and helped fund preparation of the manuscript. Thanks to Susan Perkins for her great work in putting the manuscript in a useful format. At the University Press of New England, we thank our editor David Caffry for his help in bringing this project to a swift publication. Trombulak thanks his wife Dorothy and his children Emily and Ian for their patience and support throughout the preparation of the manuscript. Klyza thanks his wife Sheila for her help and support and Middlebury College for leave support during which he was able to work on the manuscript.

We dedicate this book to all of you who read it and decide to make this region a better place, where humans live with pride and respect alongside the rest of nature.

The Future of the Northern Forest

Introduction

Christopher McGrory Klyza
Stephen C. Trombulak

Stretching from the shores of the Atlantic Ocean in Maine to the shores of Lake Ontario in New York, the Northern Forest of Maine, New Hampshire, Vermont, and New York encompasses more than 25 million acres. The region includes mountain ranges such as the White Mountains, the Green Mountains, and the Adirondacks, as well as the tallest peaks in each of the four states. The Northern Forest contains abundant wild rivers and lakes, wildlife, and wetlands. It is the home of such important recreational features as the Appalachian Trail, the Long Trail, Adirondack State Park, Baxter State Park, and the Allagash Wilderness Waterway. Yet this region is not an area without human activity. Rather, humans live and work throughout it. This characteristic distinguishes the Northern Forest from many similar forest regions in the western United States. There, humans tend to be concentrated in small parcels of private lands, surrounded by public lands and, in places, large wilderness areas. In contrast, in the Northern Forest over 80% of the land is privately owned. Nonetheless, until recently the forested ecosystems of the region had remained fairly healthy and humans and nature seemed to coexist well.

In the mid-1980s, though, a spate of corporate takeovers in the forest products industry, increased land sales, and a growing focus on forest practices prompted public concern over the status and future of the area. In 1988, the four states and the federal government initiated two parallel studies: the Governors' Task Force on Northern Forest Lands and the Northern Forest Lands Study. These studies focused on what effect these changes were having on the Northern Forest and the communities in it.

Upon completion of these studies, the states and federal government agreed to establish the Northern Forest Lands Council to study and develop alternatives to help protect the Northern Forest and its human communities. This council is still at work, having recently issued draft recommendations. It will present its final recommendations to Congress and the public in the fall of 1994.

The issues at stake in the Northern Forest policy process are complex and intertwined, as is the case with most environmental policy issues. In general, the issues involve economic, ecological, ethical, and political questions. Since the major economic activites in the Northern Forest are forest products, recreation, and tourism, what the Northern Forest looks like in the future is going to greatly affect the economic status of the region. For instance, what effect will decisions on modernizing mills have on the future of employment in the region? How will companies respond to increased regulation of forest practices? What effect will increased clearcutting have on tourism and recreation? What effect will further second home developments have on the costs of owning forest lands and on tax burdens? What would happen to the forest products industry if millions of acres of currently private land were purchased by the federal government?

From an ecological perspective, the future of a healthy Northern Forest is an emerging issue. In some parts of the Northern Forest, from the mid-1800s through the 1970s the forest of the area returned as agricultural lands were abandoned. In other areas, industrial and nonindustrial timber owners conducted their operations in a manner conducive to a healthy forest. Threats have developed in some portions of the Northern Forest since the 1970s, however, as forest practices changed and as management became more intensive. The 1980s also ushered in an increased concern over the threat of forest fragmentation and ecosystem modification due to a variety of factors, including development and some forestry practices. These changes represent a challenge to the current status of the forest, and its future as a potential recovery zone for extirpated and rare species.

Unfortunately, the public dialog on these issues has the potential to be framed to present a choice between a healthy economy and a healthy environment. We feel that this is both simplistic and unnecessary. The economy and the environment must not be viewed in an antagonistic way; rather, we must try to develop alternatives that allow for a healthy economy, broadly defined, in a healthy environment.

A number of ethical questions stand tall in the Northern Forest. How does the United States deal with the claims, desires, and needs of the Abenaki people in the Northern Forest? Do we treat these native peoples with respect, or dismiss them as no longer being relevant? Since over 80% of the land in the Northern Forest is privately owned, the issue of the rights and responsibilities of private property owners arises. What can society require someone to do on his or her land? What can society require a corporation to do? Is a private landowner without responsibilities to the community at large? Finally, the Northern Forest provides another opportunity for us to wrestle with the difficult question of how humans and nature should interact. How do we determine the trade-off of greater material wealth for humans versus restoring the wolf to the Northern Forest? What of human needs for electricity and the needs of other species for habitat? How these questions are answered in the Northern Forest will not settle these larger questions, but they will provide pieces to this large puzzle.

Finally, there is the chief political issue of the Northern Forest policy debate: Will differences among the forest products industry, environmentalists, and property rights advocates lead to political gridlock, or will a viable resolution to the future of the Northern Forest be achieved? This political issue is one shared with other regional environmental issues, such as the spotted owl protection in the Northwest and wilderness designation in Montana. If the Northern Forest policy process leads to acceptable outcomes, this process might serve as a model for the rest of the country. It could be an extremely important model because the Northern Forest covers a large area, it involves both state and federal governments, it focuses on both economics and environment, and it addresses an area with predominantly private land ownership. Success in the Northern Forest could provide a model for sustainable human and natural communities throughout the nation. Continued gridlock would further suggest that the political system is not working on these crucial and complex issues.

This book is intended to highlight the conversation on the future of the Northern Forest. It is designed to bring the reader into this maze of concerns and questions. It will introduce the reader to the major issues and actors involved; provide a context of the natural history, politics, and economics of the Northern Forest; and present a variety of perspectives on the future of the region by *participants in the policy process*. This book features both academics and political actors. The participants do not seek to analyze the issues, but rather to make the case for their visions. As such, this book differs from a more academic-oriented analysis of the issues, actors, and policy alternatives. We think that this format is more valuable because it allows the various political players to make their own case, and seek to convince the reader as to the superiority of their vision. In this book we do not weight one perspective against the next and then recommend one as optimal. Neither does this book represent a single vision that all the authors accept; each author presents his or her own argument, and is not responsible for the ideas presented in other chapters. We present these visions so that those interested in the Northern Forest can become informed citizens in the ongoing policy process. We urge the reader to evaluate these different approaches, and then to become involved in the policy process in support of the Northern Forest they would most like to see.

The book begins with five background chapters on various aspects of the Northern Forest: the natural history of the region; a native view of the region by the Abenaki people, who have called the Northern Forest home for many thousands of years; a discussion of how the Northern Forest issue arose on the agenda, the major actors involved, and the alternatives presented on the issue; an introduction to the economy of the Northern Forest; and a discussion of ethical concerns to be considered in the Northern Forest conversation.

The second section of the book contains chapters examining Northern Forest concerns and the policy process from three different public sector perspectives. First is one insider's view of the political process of the Governors' Task Force and the Northern Forest Lands Study. This is followed by a chapter by the chair of the Adirondack Park Agency (which oversees the 6 million acre Adirondack Park, comprising most of the Northern Forest in New York state), which explains how a "green line" approach actually works; and a chapter by a representative of local communities, in which the effects of any changes or new policies in the Northern Forest will be most strongly felt.

The third section of the book contains four chapters in which actors deeply involved in the Northern Forest debate present their positions. Generally, those most actively involved have been the forest products industry and environmental groups. Hence, two representatives from each of these broad groupings have contributed a chapter. Environmentalists are represented by Emily Bateson of the Conservation Law Foundation and by Jamie Sayen, editor of the *Northern Forest Forum*. The forest products industry is represented by Jonathan Wood of Bell-Gates Lumber Corporation in Jeffersonville, Vermont, and Henry Swan of Wagner Woodlands, Inc., in Lyme, New Hampshire.

The final two chapters constitute part IV. John Elder's essay on the changing New England landscape and how humans relate to it is a piece inviting contemplation on the Northern Forest. The final chapter offers our conclusions on the Northern Forest conversation and the direction the region is taking. We seek to underline areas of common ground and discover the basis for disagreements; to stress the most important components of the region—culturally, ecologically, economically, and politically; and to offer our view of where the debate ought to head in the Northern Forest.

Since the ecological boundaries of the Northern Forest do not correspond to political boundaries, we had at the outset sought a contribution examining the issue from the Canadian perspective. Over the last year, we held discussions with a number of people from Canada: academics, activists, government officials, forest products representatives. Based on these conversations, it became clear that the Northern Forest issue does not have the same salience in New Brunswick and Quebec as in New England. Environmentalists have become increasingly concerned about forest practices in the region, and throughout Canada, but the focus of their concern is not on changing ownership patterns. This is most likely due to the land ownership pattern in these two provinces. In New Brunswick, 49% of the forest land is Provincial and Federal Crown lands (analogous to public lands in the United States). Similarly, in Quebec 88% of inventoried productive forest land is Provincial Crown land. This degree of public ownership serves to cushion the blow of any changes in the private sector.

The outcome of this conversation about the future of the Northern Forest is important at two levels. It is important for this 26-million-acre region, because decisions made and policies adopted or rejected will significantly affect the trees, people, rivers, and animals that call the Northern Forest home. But also, the results of this discussion could present a model as to the future of humans living in nature. In the Northern Forest perhaps we can chart a future of humans living with nature rather than against it, of truly sustainable human habitation. Perhaps, too, this conversation could demonstrate a way for people with diverse opinions to find common ground and move forward when dealing with difficult issues. We hope that this book helps contribute to the development of sustainable human and natural communities in the Northern Forest and beyond.

PART I

Understanding the Northern Forest Today

Before the current political debate over the future of the Northern Forest can be explored, it is important to discuss the basic issues that have shaped the history of this region and the people that live here. The most important of these issues are ecological, cultural, political, economic, and ethical. We have asked eight authors—including ourselves, the two editors of this book—to address these topics in a way to help the reader understand the Northern Forest region today.

The debate over the future of the Northern Forest is essentially a debate over the relationship between nature and human traditions. The first step in understanding this debate, therefore, is an appreciation for the natural history of this area. In the first chapter, Stephen C. Trombulak, a field ecologist at Middlebury College, uses his experience as a conservation biologist to describe the history of the natural forest development, the forces—both natural and human—that have continuously shaped these forests, and the consequences of disturbance to the biotic integrity of this region. He has been closely involved with the environmental aspects of the Northern Forest debate, being a member of the Board of Directors of the Vermont Natural Resources Council, contributing editor of the *Northern Forest Forum*, and a member of the Workgroup for the Biological Resources Diversity Subcommittee of the Northern Forest Lands Council.

The Northern Forest issue is not just about ecosystems, however. It is about how human societies exist within these ecosystems. Therefore, we focus intensively on the human dimension of this issue. Although the dominant human culture today derives from European colonization begun over 300 years ago, it was not the first culture to thrive here, nor is it the only culture to live in and depend on these forests today. A proper understanding of the future of humans in the Northern Forest must begin with an understanding of the first cultures, those of Native Americans. Although several different Native American cultures have inhabited this region in sequence since the retreat of the glaciers 10,000 years ago, and several different tribes currently claim portions of the Northern Forest region as their ancestral homes, the perspective offered by any one tribe well exemplifies the attitudes toward land, nature, sovereignty, and rights held by the others. Four speakers for the Abenaki Tribal Council—Hilda Robtoy, Dee Brightstar, Tom Obomsawin, and John Moody—give us this perspective. Each of them has long been an outspoken advocate for Abenaki rights and preservation of cultural heritage. Recognizing this important element of human culture in the region is essential to shaping a just future for all human societies.

The area referred to as the Northern Forest is, however, not just one ecosystem, forest type, or tribal homeland. It is most specifically a political entity created by the U.S. Congress to address a host of concerns of the people in specific portions of New York, Vermont, New Hampshire, and Maine. Christopher McGrory Klyza provides background on the political dimensions of this issue. Klyza is a political scientist at Middlebury College with expertise in natural resources policy. He has written widely on natural resources issues, especially those concerning public lands.

Thomas Carr explores the current patterns of economic activity in the Northern Forest, not only for the timber products industry—the economic sector that has driven much of the political discussion on this issue so far—but for the recreational and agricultural sectors as well. Carr is an environmental economist at Middlebury College with particular expertise on issues associated with resource-based economies.

The final human dimension to this debate is one of human ethics: What are the ethical tensions among different human communities in this region? between human and natural communities? Stephanie Kaza, a professor of environmental studies in the Environmental Program at the University of Vermont, focuses her teaching and writing on the religious and ethical dimensions of environmental issues, including a recent book on the relations of humans and forests.

These chapters will allow the reader to understand the perspectives offered by the authors in the subsequent parts of this book, especially to judge how well any of these authors understand (1) all of the critical issues that face the Northern Forest, (2) the essential elements to any lasting solutions that improve the future for the people and ecosystems of this region, and (3) the steps that will be necessary to develop these solutions. In a sense, the five chapters in this part help the reader become more knowledgeable of the fundamental characteristics of this region and the bases for the debate. Readers can then act not as passive consumers of the visions and opinions of others, but as informed and active arbiters of successful solutions.

1

A Natural History of the Northern Forest

Stephen C. Trombulak

Fifteen thousand years ago the land we call the Northern Forest lay under a sheet of ice more than a mile thick. The great continental glaciers had come and gone several times before, and were once again retreating from their southernmost reach near Long Island. As the climate warmed, the leading edge of the ice sheet retreated further north, and by 11,000 years ago the northern portions of what would become the states of New York, Vermont, New Hampshire, and Maine were once again uncovered.[1] Quickly the bare rock and glacial till were colonized by cold-adapted plants, fungi, lichens, and bacteria, and in less than 1,000 years much of the surface rock had been converted to soil, trees had migrated north from their southern refuges, and forest communities became once again the dominant ecosystem of the region.

The story of the Northern Forest is above all else a story of nature. The political, economic, and cultural challenges that we face today are, and always have been, played out on a stage dominated by the earth, trees, climate, water, and animals. Therefore, to understand why so many people are concerned for the future of the Northern Forest and what the natural constraints on visions for its future are, we must first understand what the Northern Forest is from an ecological perspective.

Discussing the natural history of the Northern Forest unavoidably results in complex and diverse descriptions because, despite political attempts to deal with the region as if it were ecologically uniform, the Northern Forest includes more than one ecoregion, each with its own patterns of biodiversity, soil, and topography. The Environmental Protection Agency, for example, describes the Northern Forest as including portions of the Northeastern Highlands, Northeastern Coastal Zone, and Northern Appalachian Plateau and Uplands ecoregions, each of which also has an extensive distribution outside of the Northern Forest.[2] The U.S. Fish and Wildlife Service places the Northern Forest at the intersection of three physiographic regions: the Adirondack Mountains, Northern New England, and Northern Spruce–Hardwoods.[3] I attempt here, however, to minimize the problem of such ecological diversity by emphasizing the characteristics of the region that are common throughout, and to treat regional differences as if they were equivalent to the normal patterns of variation seen within any single ecoregion. This approach bows to the political realities of the current debate over the future of the region but will, I hope, signal that simple descriptions and single solutions for the entire region are unrealistic.

In this chapter I focus on three key questions concerning the natural history of the Northern Forest, questions that set the stage for all the chapters to follow. First, what natural features of this region make it unique? Second, what are the natural ecosystems that have become established in this region? Third, what impact has human occupancy had on these ecosystems? I end with observations on ways in which a scientific perspective can contribute to discussions of the Northern Forest's future and help us to determine which of the many competing visions for the future can realistically be implemented, given the constraints of the natural features of the region.

What Natural Features of This Region Make It Unique?

To many, the Northern Forest is special simply because of its size and location. The officially designated region comprises over 26 million acres of essentially forested land stretching for over 450 miles from the northern tip of Maine to the Tug Hill region in upstate New York (Figure 1.1).[4] Forests of the types that characterize this region also extend north and south to include several million additional acres, but have been excluded from the Northern Forest for political rather than ecological reasons. Immediately surrounding the Northern Forest are the biologically identical forests of the southern Green Mountains of Vermont, the Berkshire Mountains of western Massachusetts, the southern White Mountains of New Hampshire, the Catskill Mountains of New York, and the southern boreal forests of Quebec, New Brunswick, and Nova Scotia in Canada. The Northern Forest has also been identified by geographers in the U.S. Forest Service as being part of the larger Laurentian Mixed Forest Province that stretches from Maine to Minnesota.[5]

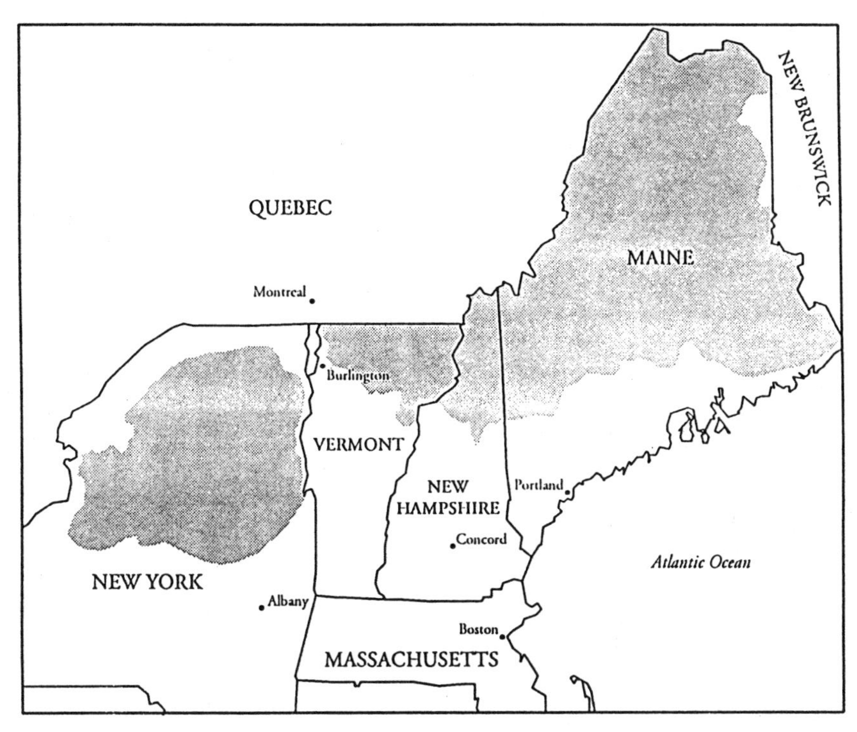

Figure 1.1. The Northern Forest of New England and northern New York, as defined by the U.S. Forest Service and the Governors' Task Force. Shown in solid lines are the states of Maine, New Hampshire, Vermont, and northeastern New York. The shaded area marks the location of the Northern Forest. In New York this area encompasses the Adirondack Park–Tug Hill region. In Vermont, New Hampshire, and Maine the northern boundary is the United States–Canada boundary. (Adapted from Harper, Stephen C., Laura L. Falk, and Edward W. Rankin, 1990, *The Northern Forest Lands Study of New England and New York*, Rutland, Vermont: U.S. Department of Agriculture, Forest Service; Figure 1, p. x.)

Size alone does not make this area unique. Many regions of ecological interest in western North America are comparable in sheer acreage, such as the Basin and Range Ecosystem of Nevada and the Arctic/Yukon Flats National Wildlife Refuges in northeastern Alaska. These western areas, however, are in many ways less connected to or influenced by human development. Few people visit them; fewer still live and work in them. The Northern Forest is different. Situated in the northeastern United States it has had a long history of human occupancy. Since the 1600s it has not conformed to the definition of "wilderness" that has grown out of western environmental tradition: nature untouched by humanity.

Paleo-Indians probably moved into this area very quickly after the glaciers receded 12,000 years ago.[6] Their occupancy required an intimate association with the forest and use of its flora and fauna. Although evidence suggests that the various Indian cultures that developed in the Northern Forest did not practice the extensive forest burning and clearing of those who lived in southern New England,[7] they were nonetheless widespread throughout the area, with a total peak population of perhaps 20,000 concentrated in seasonally shifting villages.[8]

The transition from a dominant Indian culture to a dominant European culture was very rapid. European settlers arrived in southern New England as early as the 1620s[9] and spread steadily northward over the following 70 years. Today, about 1 million people live in this region, with an additional 70 million people, because of the interstate highway system, living within an 8-hour drive.[10] This establishes the Northern Forest region as a premier natural recreational area for almost 30% of the population of the United States, as well as for the population of the major urban centers of southeastern Canada, with extensive opportunities for camping, hunting, fishing, hiking, canoeing, and skiing, and just plain "getting away" to nature.

In short, the Northern Forest has had over 10,000 years of history of occupancy by humans who depended, and continue to depend, on many aspects of the forests, particularly its trees and animals, to support their existence. Simultaneously, as a result of patterns of land use and human settlement throughout the United States, the Northern Forest is the largest expanse of forest ecosystem near major population centers, and as such is a unique natural resource for much of the country.

What Natural Ecosystems Have Become Established in This Region?

When the glaciers receded, they left a land of varied topography. In the east are low-relief plains where peaks rarely rise more than 1,500 feet above the surrounding lowlands. In the west are the Adirondack Mountains, the product of several separate events of continental collision and uplift begun over 900 million years ago, carved and polished but otherwise unchanged by the glaciers. In between is a system of valleys and ranges that run along north–south axes: Champlain Basin, Green Mountains, Connecticut River Valley, White Mountains, Kennebec River Valley, and Penobscot River Valley.

A tremendous diversity in ecosystems is found throughout the region, due to this wide range of topography. In general, the structure and function of ecosystems are brought about by the interplay of both biological components and features of the physical environment, including climate, soil, water, air, and energy from the sun.

The climate in this region is best characterized by warm, wet summers and cold, snowy winters. The combination of the region's northern latitude and proximity to the eastern edge of a continent drives a pattern of wind that brings cold arctic air down from the north in the winter as storms move in from the west, and warm wet air from the midwest and the Gulf of Mexico during the summer. As a result, abundant precipitation falls throughout the year, with large amounts delivered both as snow and rain.[11]

The soils of the Northern Forest vary over the region and both influence and are influenced by the forests that grow in them. In many places, especially the higher elevations in the western portion of the region, soils are thin or nonexistent, resulting in large areas of exposed bedrock. In general, the soils throughout the region are typical for forests at high latitudes: generally shallow and well drained, composed of stony or sandy loams, and developed in till left by the retreating glacier. These soils are acidic, due in part to the short growing season, which promotes the buildup of incompletely decayed organic matter, and in part to the presence of conifers like spruce, fir, and pine, which acidify the soils in which they grow.[12] These characteristics make the soil generally poor for agriculture and highly fragile. Coupled with the elevational relief in some parts of the region, the potential for soil erosion is great, especially when protective ground cover is removed.

Climate and soil characteristics are the dominant factors that determine the biological components of the ecosystems in this region. Perhaps the single most important natural aspect of the Northern Forest is the forests themselves. Two basic types of forests dominate the region: the spruce–fir forests and the northern hardwood (or maple–beech–birch) forests.[13]

The spruce–fir forest is today the most common type in northern Maine and the upper elevations of New Hampshire, Vermont, and New York. Several species of coniferous trees belong to this community, particularly red spruce, black spruce, white spruce, balsam fir, eastern larch, and northern white cedar. This forest type is typically found in areas with cold temperatures and acidic soils, although, as I discuss later, other forest types readily intergrade with this community or become established in areas where the spruce–fir forests are heavily disturbed.

The northern hardwood (or maple–beech–birch) forest is the most common throughout the New Hampshire, Vermont, and New York portion of the Northern Forest, and is found as well in a few regions of Maine. This forest community is dominated by sugar maple, American beech, yellow birch, red maple, and black cherry. It is tolerant of a wide range of conditions, but is usually found in areas that are relatively warm, such as are found in the more southern latitudes of the Northern Forest and at lower elevations. In many places, spruce, fir, and northern hardwood species grow alongside one another, demonstrating that these forests are not unvarying units but general categories that commonly intergrade with one another.

Three other forest types are present to a much lesser degree. Aspen–birch forests, comprised of quaking aspen, bigtooth aspen, paper birch, balsam poplar, and pin cherry, are found in a few locations throughout the region. This is an early-successional forest type that is found only in areas of recent disturbance, either natural or human-caused. Left undisturbed, these forests succeed to either spruce–fir or northern hardwood forests, depending on the location. White–red–jack pine forests, composed of those pines as well as eastern hemlock, red maple, chestnut oak, and northern red oak, are found sporadically in Vermont and New York. Although each of these pines is present in the region, of most importance historically were the eastern white pines, which at one time were the largest of all conifers in the region. Trees up to six feet in diameter and 200 feet tall were common, but were quickly removed due to their importance as masts for the ships being built in England and Boston.[14] Oak–hickory forests are restricted to just a few sites in the Adirondack Mountains. These stands represent the northernmost distribution of this predominantly southern forest type.

The forests in this region have never been static entities. Changes in forest structure and composition have been driven by two dominant forces. The first is the natural process of succession following disturbance unrelated to human activities. In some respects, the entire development of the Northern Forest can be thought of as the successional establishment of mature forests following the disturbance of the last ice age. After the glaciers receded, the first trees to colonize the newly exposed tundralike area were spruces, followed shortly by paper birch, alder, and balsam fir.[15] These tree species were replaced over the succeeding millennia by more southern softwood and hardwood species, first white pine, birch, and oak, then later by hemlock and beech. The relative frequency of these species varied considerably through time and in different parts of the region; comparing vegetation assemblages at any one location over time or among locations at one time shows little evidence for stability and uniformity in forest structure. Indeed, the forest types seen in the region today took on their present configuration only between 3,000 and 1,000 years ago, as an apparent cooling trend brought spruce and fir southward again to become established in the more northern latitudes of Maine and at upper elevations in the mountains of New Hampshire, Vermont, and New York.

Sequences of change in vegetational structure following disturbance occur on smaller scales as well. For example, following fire or extensive storm damage in the spruce–fir forest, the area is first colonized by aspen and birch, which are more tolerant of light and therefore grow better in newly cleared areas than the more shade tolerant climax species. Reconstruction of forest history in Maine, for example, suggests that at any particular site major fires burned about once every 800 years and that hurricanes caused major storm damage once every 1,000 years.[16]

Other natural disturbances occur that cause less of a change in the species composition of the forest and more of a change in the physical structure. Waves, or lines, of balsam fir death and regeneration can be seen on mountain sides throughout the Northern Forest where winds cause blowdowns of old firs that are exposed at the leading edge of a wave. After the trees die they are quickly replaced by fir saplings, which continue to grow until the next wave edge reaches them in 60 to 70 years and they once again are blown down.[17] Other natural disturbances can also remove trees from large areas of forest, including large-scale attacks by insects.

All of this implies that, although forest structure can come to exemplify a characteristic climax community, natural forest ecosystems do not long exist in some kind of equilibrium state either in local or broad regional areas. Clearly the "natural" composition and structure of a forest are influenced by many factors, particularly climate and disturbance, all of which operate on their own scale over time and space.

This also implies, however, that when rates and scales of disturbance are low and rates of environmental change are slow, forest communities tend toward a climax stage characteristic for a particular climate and set of available species. This appears to have been the situation prior to the colonization of the region by Europeans. Before large-scale timber harvesting began, forests tended to be old, trees were large, and climax communities were characteristic in large portions of the region.

The second force that has shaped the ecosystems of the Northern Forest has been human disturbance, primarily from clearing for agriculture and settlement and from commercial timber harvesting. The effects of these disturbances will be discussed more fully in the next section; it is clear, however, that this force has a different effect on the natural ecosystems of the area than the disturbances discussed earlier, because of their frequency, magnitude, and duration.

Despite the region's name, it would be a mistake to imagine that forests are the only ecosystems found here. Also present are 68,500 miles of rivers and streams, over 1 million acres of lakes, and over 2½ million acres of wetlands, all of which play vital roles as plant and wildlife habitat, as parts of the local hydrological cycle, and in erosion control.[18] Scattered alpine and subalpine communities exist at higher elevations on the taller mountains in New York, Vermont, and New Hampshire, retaining species compositions characteristic of the original communities to colonize the Northern Forest after the retreat of the glaciers.[19] All of these ecosystems are functionally linked with those of the forests through the exchange of water and nutrients as well as their shared importance to the animals that live in the region, and create a network of diversity that defines the biotic structure of the region.

It would also be a mistake to think of forest ecosystems as being solely collections of trees. The complex biotic relationships within an ecosystem involve more than just the dominant tree species. We have no idea how many different species of organisms live in the Northern Forest region. Some groups are quite well known. For example, we know with a high degree of confidence that there are about 400 species of birds, 55 species of mammals, 11 species of reptiles, and 19 species of amphibians found in the Northern Forest.[20] Other groups are less well known, such as vascular plants, fungi, and insects. Some groups, like soil bacteria, are barely known at all.

The extent to which a group has been cataloged, however, is not indicative of its role in the ecosystem. Ecologists are quite certain that few parts of an ecosystem are superfluous, and that trying to characterize and manage ecosystems based on presumed "dominant" species is inherently impossible.[21] Bacteria and fungi decompose dead tissue, allowing essential nutrients to be taken up once again by plants at the base of the food chain. They also play a vital role in the creation of soil; the natural acids they produce begin the process of chemical weathering of rock that eventually results in the rock particles and minerals becoming physically mixed with organic matter to form top soil. Soil invertebrates, like earthworms, are also involved in this process, mixing the soil through the action of their burrowing, bringing nutrients up from the depths, allowing oxygen to enter from above, and generally breaking up the soil to allow water to flow through. Understory plants, like ferns, club mosses, and other small plants, protect the soil from erosion. Predatory insects and birds exert biological control on insects that eat plants. Amphibians help transfer nutrients and energy across aquatic and terrestrial habitats. Deer, moose, and other mammalian herbivores cycle nutrients through an ecosystem. Large carnivores keep populations of many species healthy and from growing beyond the carrying capacity of the environment.

Other aspects of the organisms in the Northern Forest can only be appreciated by looking at the region in a broader context. For example, some of the species found in the Northern Forest are essentially found nowhere else. All together, 25 species of vertebrates are unique to the Northern Forest.[22] This means that anything that removes one of these species from this region removes it permanently from the earth.

Also, the Northern Forest is linked to other major biomes. It is part of the Atlantic flyway for migratory waterfowl moving between their breeding grounds in northeastern Canada and their wintering grounds in the south. Even more critical is the connection between the Northern Forest and tropical forests. Forest-dwelling, insect-feeding birds that breed during the summer in North American forests, including numerous species of warblers and vireos, spend their winters in tropical forests of the Caribbean, Central America, and northern South America. Anything that affects the birds in one of these areas will be felt in the other.

What Impact Have Humans Had on Natural Ecosystems in This Region?

Although humans have lived in the Northern Forest for millennia, the major impacts to the ecosystems of this area came with European colonization after the 1600s.[23] The original expanse of old growth was eventually cut over as the forests were selectively cut or cleared for agriculture and the trees harvested for fuel, timber, and, in the case of white pine, masts for ships. The transformation of the forests did not occur suddenly, but over the course of about 200 years the old, massive spruce–fir and northern hardwood forests were cut. It is estimated that by the 1850s over 70% of Vermont's original forests had been logged off.[24] In Maine, over 30% was cut by the middle of the 1800s, and virtually all of it had been cut by the end of that century.[25] Forests later reestablished themselves throughout much of the area. For example, forests are now found over 75% of Vermont,[26] a testimony to the regenerative powers of these forests. Some areas of the Northern Forest have actually experienced second and third waves of cutting with subsequent regeneration.

These replacement forests differ from the primary forests in a number of important ways, however. The trees are smaller, younger, and denser than was the case in the original old growth. They also, by and large, represent forests that are at earlier stages of succession. These differences can result in changes in soil conditions, temperature, and water availability, which all have impacts on other aspects of the forest ecosystem. Duffy and Meier, studying hardwood forests in the southern Appalachian forests, demonstrated that species composition of the herbaceous understory of these forests still had not returned to the state seen in undisturbed old growth even 87 years after cutting.[27] This strongly suggests that successional processes that lead to climax forest structure take place over centuries rather than decades.

It should be recognized that an impact on forest structure is felt by these ecosystems regardless of whether the forest is cleared for development—the form of ecosystem disturbance that first drew attention to the Northern Forest—or for timber harvesting, which is currently the dominant type of human influence over most of the region. Each will have different effects, however. Land transformed to housing developments of some type will almost certainly not follow a successional path back to forest. The clearing, paving, water diversion, and impacts of human occupancy, such as the production of sewage, solid waste, and air pollution, all have lasting impacts on natural ecosystems.

Timber harvesting also influences forest ecosystems, but the influences are highly dependent on how the harvesting is done. Small-scale harvesting that (a) minimizes the use of heavy machinery, (b) is carried out during the winter, and (c) retains the mixed species composition and uneven-aged stands of the natural forest is likely to have minimal impact on patterns and processes in these ecosystems. The ecological consequences of timber harvesting increase, however, as harvesting practices (a) become more energy-intensive, (b) are carried out during seasons when the ground is wet, (c) harvest a larger percentage of the trees in an area, (d) are carried out over a larger area, (e) involve more road construction, and (f) alter more of the local conditions after cutting.

It is not my purpose here to discuss the practices of particular companies or the effects their operations have on the long-term health of the forest ecosystems. I only intend to point out the effects of particular practices so that proposals made in subsequent chapters or at any point in the ongoing debate over the future of this region can be placed in the context of their ecological consequences. The following list highlights harvesting practices employed in some parts of the Northern Forest, and the impacts they have on natural ecosystems.

1. *Cutting and removing trees from large areas.* This type of tree removal does not resemble any natural type of disturbance. In contrast to fire, storms, and insect damage, large-scale clearing removes most of the plant-based nutrients from the area, does not leave standing dead trees (snags) that are important nest sites for birds, and exposes more of the soil to direct sunlight.

2. *Use of heavy machinery in harvesting operations.* Whole-tree harvesters, bulldozers, and loaders heavily compact the soil and, if the soil is wet, cause extensive rutting.[28] Road construction associated with the use of machinery also substantially increases erosion. These impacts affect the ability of forests to regenerate after cutting. It is also likely that they alter the balance of soil microbes, fungi, and invertebrates that influence patterns of nutrient cycling during successional processes.[29]

3. *Herbicide spraying to reduce unwanted tree species.* This practice alters the natural species composition of the forest as well as patterns of succession and soil replenishment. These herbicides are also transported to other locations within the ecosystem, getting into surface water, and

potentially harming both terrestrial and aquatic organisms. For example, in Maine herbicide spraying is used to eliminate hardwoods and increase the volume of commercially more desirable spruce. Removal of the hardwoods has resulted in an increase in the density of balsam fir, a preferred food for the spruce budworm, a major insect defoliator of softwoods, and as a result has increased the severity of spruce budworm outbreaks throughout this century.[30]

4. *Replanting cut areas with monocultures of commercially desirable species.* Monocultures of any species are rarely natural, and often lead to the breakdown of natural ecosystem processes. For example, monocultures are generally more susceptible to insect damage because these artificial ecosystems do not have the natural complement of insect predators or the suite of plants that protect each other through their complementary anti-herbivore strategies. Also, agricultural (including sylvicultural) practices that emphasize the establishment of monocultures can introduce or spread species and genetic strains that are not native to, or are normally uncommon in, an area. This is true in many areas of northern Maine, where red spruce forests that have been cut are replanted with white or black spruce, species that are normally restricted to wetter sites in this region.

5. *Repeated cutting in areas at short intervals.* As mentioned before, the frequency of large-scale natural disturbances throughout most of the Northern Forest was on the order of centuries, a time scale that allowed soil to be replenished and forest types at later stages of succession to become established. When forests are cut at intervals of only decades, however, these processes are inhibited, natural ecosystems are not reestablished, and species that are dependent on these later stages are threatened.

The impact of human disturbance is also felt in areas that themselves are not directly subject to disturbance. In particular, the effects of cutting are felt in portions of forests that are left standing but are in proximity to areas that are cut. Large-scale cutting in the Northern Forest has not been restricted to only a few areas within the 26 million acres, nor has it occurred as a single wave of disturbance that has moved evenly across the landscape. Instead it has happened throughout the region, leaving patches of forest in a background of cut or intensively managed land. These patches themselves experience altered conditions that affect the patterns and processes that would be found in an undisturbed forest.

The effects of fragmentation are of two types.[31] The first is simply the reduction of the total amount of area of undisturbed forest that is available to forest inhabitants. Individuals of all species have a minimum area required to carry out the activities that are required by their life history, such as acquiring food and reproduction. Fragmentation reduces the size of the undisturbed area and therefore decreases the number of species that can persist. The first to disappear from an area as a result of fragmentation are, of course, those species with the largest home ranges, such as the large mammalian carnivores like mountain lions, wolves, and lynx, which may have home ranges of many tens of square miles.

The second effect of fragmentation is the modification of conditions in the fragment as a result of its proximity to disturbance. These edge effects occur due to alteration of physical conditions, such as changes in wind and temperature, as well as changes in the biological characteristics near the border of the patch. For example, forests close to areas of disturbance have a higher proportion of weedy grasses and browse shrubs than do forests further away. Similarly, species that are found predominantly in disturbed areas, such as raccoons, blue jays, and deer, are more common in forests near disturbances than they are away from them.

How large a fragment must be to prevent portions from experiencing edge effects depends on the organism and shape of the fragment. For example, nest predation on forest-dwelling birds from predators that enter from disturbed areas is felt up to 600 meters into a forest from the nearest edge.[32] If a fragment is exactly 1 square kilometer (1000 meters on each side), then no part of the fragment is more than 500 meters from the nearest edge, and as far as birds are concerned the entire block of forest is disturbed habitat. A circular fragment would have to be more than 1.1 km^2 before any portion is more than 600 meters from the nearest edge. This minimum size, however, refers only to a circular fragment; a fragment could be 100 km^2, for example, and still comprise nothing but disturbed edge if it were long and thin, such as 100 km by 1 km.

The effects of fragmentation on forest ecosystems are numerous. Species that cannot tolerate disturbed habitat and whose home ranges are large or that require a wide diversity of habitats are likely to go locally extinct. Species that favor habitats or conditions found in disturbed or open areas will increase. Deer, woodchucks, and raccoons, for example, have increased tremendously in response to the opening of forests for development and intensive timber management, as have a few bird species.[33] Species that are sensitive to disturbance or are negatively affected by the changes brought by the forest clearing will decline. The best documented case of this type has been of the forest-dwelling insectivorous migrant birds. It has long been noted that forest songbirds were declining throughout the eastern and central United States. Because these birds were generally migrants, it was assumed that the decreases in their numbers were caused by destruction of the tropical forests that are their wintering grounds. However, recent studies have demonstrated that migratory patterns do not correlate with the degree of decline; nonmigrants are just as likely to decrease as are tropical migrants. The data suggest instead that the declines are due to predation in the temperate forest breeding grounds from species such as raccoons, dogs, and cats that are now active in forests due to their presence in disturbed areas that border these fragments.[34]

The effects of fragmentation on forest ecosystems are highly dependent on the type of disturbance that surrounds the undisturbed land. The animals that live in mature forest stands surrounded by urban development experience very different conditions than those that live in stands surrounded by young, regenerating forests. Analyses of the impacts of fragmentation on forest ecosystems must be careful to evaluate the response of organisms to different kinds of disturbance.

This diversity of responses to fragmentation—population increases for some species and decreases for others—demonstrates the difficulty of making simple statements about the effect of land-use practices on forest ecosystems. One must instead look at which species are increasing and decreasing and assess whether those species are natural members of the ecosystem and whether they are threatened or endangered throughout their range.

The problems that result from fragmentation can be solved or prevented only by minimizing this type of disturbance. Parcels of undisturbed forest must not only be allowed to persist, but they should be large (many square miles at a minimum to support populations of carnivores) and symmetrical (to minimize the amount of disturbed "edge" relative to undisturbed "core" area), encompass diverse habitat types, and follow natural features that are important for animal and plant movement, such as rivers. Also, considering the history of long-term patterns of geographic range shifts in the Northern Forest region, parcels should exhibit a high degree of interconnectedness. Connections among parcels that allow for the movement of species in response to natural or human-caused changes in climate may play a critical role in allowing species to reestablish themselves and avoid local extinction.

Human occupancy has had other effects on the integrity of the forest ecosystems of the Northern Forest region. Historically, hunting was responsible for the local extinction of a number of mammals and birds, including the timber wolf, mountain lion (or catamount), pine marten, fisher, wolverine, elk, beaver, peregrine falcon, and turkey.[35] All told, almost 100 species of plants and animals are currently listed as threatened or endangered in this region; many others probably ought to be listed but are not due to our incomplete knowledge of the region's biota. The decrease or loss of any component of an ecosystem could have potentially serious consequences for ecosystem stability, especially if the species plays a critical role in the control of herbivores such as porcupine, white-tailed deer, or plant-eating insects.

The development of roads has also increased the threat to animals from automobiles. Brock has estimated, for example, that in the Adirondack Park the greatest threat to biological diversity comes not from fragmentation but from crossing the road.[36] Air pollution, particularly material that is produced from outside of the Northern Forest region, is also having an increasing impact on the health of the forests. Acid precipitation, ozone, and lead have all been shown to be altering the health of trees and the composition of the forests.[37]

Perhaps the most unappreciated effect of human occupancy on the Northern Forest has been the effect of species introductions. At first thought the introduction of a species would seem to increase the biological diversity of an area, rather than threaten it. Indeed, there are numerous examples of exotic species that have had no obvious effect on an area beyond the addition of one more species—for example, Norway spruce and brown trout. However, these examples are counterbalanced by those where the introduction of a species has led to substantial alteration of natural ecosystems. Chestnut blight, a fungus brought into the United States in the early 1900s from Asia, has all but eliminated the American chestnut. Dutch elm disease, caused by another fungus, has done the same to the several native elms. Eurasian milfoil has colonized lakes throughout the northern United States, causing substantial changes in species composition and physical characteristics of these aquatic ecosystems.[38]

In summary, the effects of human occupancy on the natural ecosystems of the Northern Forest are diverse. They vary among locations, being different in the Wild Forests of the Adirondack Park than in the industrial forests of Maine, and different in the southern boundaries near major urban centers than in less accessible northern areas. Some of this spatial variation is a function of the patterns of human occupancy, but much of it derives from the ecological heterogeneity of the region as well. The effects of human occupancy vary over time, changing in type and intensity from the time of the Paleo-Indians through the early centuries of European colonization to today, when the human population has increased over 50-fold from the time of first settlement. And the effects vary in consequence, increasing the populations of some species and decreasing others, altering the species composition of some communities and eliminating others all together.

The Role of Science in Designing a Vision of the Future

As will be made clear in the remaining chapters, most of the visions of the future of the Northern Forest center on political, economic, and cultural aspects of the issue. Often little regard is given for the role that science can play in evaluating the consequences of proposed actions. I feel strongly, however, that we will never achieve our social goals of political self-determination, economic health, and long-term cultural stability without recognizing the contributions of a scientific perspective to this debate.

The first point that must be recognized is that the scientific principles governing the operation of nature are not arbitrary, nor can they be ignored if they do not fit conveniently into desired political or economic strategies. They are not equivalent to economic or political traditions. They are the fundamental bases of our existence. Social policies either conform to principles upon which nature operates, and therefore can be sustained indefinitely, or they do not, and can only be pursued for a short time before resources are exhausted or too much capital must be spent to solve the problems the policies create. Ecological dynamics like erosion, insect pest outbreaks in managed ecosystems, loss of groundwater recharge when wetlands are filled, and population explosions of herbivores when predators are eradicated may be inconvenient and costly to avoid, but they cannot be ignored.

A scientific perspective also forces us to focus on our best available information on how ecosystems function and what might likely happen to any part when the system is disturbed. Economic and political analyses of situations usually focus solely on resources that are of particular interest, be it trees, agricultural land, minerals, or wildlife, and then only in the short term. Other factors and future consequences are labeled "externalities" and ignored in cost–benefit analyses designed to identify "optimal" strategies. Because ecosystems function as a network of interconnected parts, a scientific understanding of ecosystem processes clearly demonstrates that, for society as a whole, there are no environmental externalities! Cost–benefit analyses that focus solely on a few components of a system are unlikely to reveal strategies that are truly optimal for an entire society over time periods longer than a few years.

Of course, our understanding of any system, be it ecological, economic, or political, is never perfect, and we should continually seek to improve our knowledge of how these systems work. This does not diminish, however, the importance of using the best available scientific knowledge to articulate how we think ecological systems operate. Our less-than-perfect knowledge does not mean that science should not play a role in policy. Rather, it means that all members of society have a stake in ensuring that we use the best available scientific information in forming policy, rather than misinformation designed to achieve the goals of only a segment of society.

Finally, a scientific perspective provides a way of assessing strategies for achieving a vision of the future. Many of the chapters in this book offer personal perspectives for what the future of this region ought to be and how we can get there. Science provides not only a description of the natural world but also a way of analyzing information presented in these visions.

A positive future for the Northern Forest will come because disparate visions, such as those presented here, are compared by those involved in shaping the future to find common ground and to resolve the points of disagreement. These disagreements will be clearly resolved only when those who hold a view are forced to address three critical questions that emphasize natural laws, evidence, and probabilities: What do you think are the ecological consequences of your vision of the future? What do you think are the chances of these consequences occurring? And why do you think so? Only through critical analysis of the answers can we move beyond political and social rhetoric to achieve a sustainable future.

2

The Abenaki and the Northern Forest

Hilda Robtoy
Dee Brightstar
Tom Obomsawin
John Moody

The great forests of the ancient times are gone because of colonialism. Five hundred years ago there was a quiet on this land that had never been broken. We lived here with other Native Nations in a kind of peace that did not leave scars on this land and life. You have asked for our perspective on the "Northern Forest." We will give you a little history, then a warning and an invitation. We hope that you are able to listen, able to understand what we say.

History: Crisis in the Northern Forests

In the past few years, there has been a lot of talk about the "Northern Forest." Since the coming of the *Awanocak*/French/Canadians and the *Bastoniak*/English/Americans, the forest of our homeland has been traded on as never before. Many powerful people, so called "owners" of the land, used large parts of *Ndakinna*, Our Land, also called *Wobanaki*, the Dawn Land, for logging and other profit. Then, in the hunger for land on the lakes and rivers, fueled by the strong economy of the 1980s many of these large land "owners" began to sell off their holdings for development. A cry went up as another great cycle in the exploitation of the land began. The Northern Forest Lands Council was rapidly convened to help deal with the issue.

Since the early twentieth century, towns, states, and federal agencies have acquired major parts of the Adirondacks, Green, and White Mountains, and the lake, mountain, island, and seashore sections of Maine. They also have come into the Missisquoi, Lamoille, Winooski, Connecticut, Merrimack, Saco, Androscoggin, Kennebec, and other rivers of our homeland to acquire the wetlands. Many people have come to value the shores, the rivers, the remoter parts of the forests and mountains. Since the late 1800s, this rush to "own" a piece of the mountains and waters has become a flood. From 1600 to 1850, Euro-American logging tore down most of the ancient forests.

We must admit, in the past five hundred years, your *attention* has been simpler to *divert* than to *shape*. In the war years from 1500 to 1800 we largely kept you out of our homeland. From the 1630s, you sought council and signed many treaties of peace guaranteeing our homeland and way of life. At Pemaquid in 1693, and again at Pejebscot in 1699, Falmouth in 1703, Piscataqua in 1713/1714, and Arrowsick Island in 1717, your governments promised to respect our boundaries, trade fairly, freely accept our continued rights to fish, plant, hunt, and gather on lands we allowed you to live on. As late as 1752 the *Bastoniak* or English of Boston were given a statement from us spoken by *Atikwahondo* in which we "forbid you to kill a single beaver, or to take a single stick of timber on the lands we inhabit." Those lands stretch from Maine to Lake Champlain, the St. Lawrence to the Massachusetts foothills!

You were so afraid of the vast "wilderness," the woods and swamps, the mountains and "tractless lands," that we were able to keep you at bay with your own fear. As you moved into our homeland and forced us back, we retreated to those places you feared most, the windy islands on the lakes, the edge of swamps, the remoter, rockier parts of your towns. We sought refuge with the shrinking forest and marshes.

In the early twentieth century, all over the region our people were still the keepers of these places as game wardens and guides known to the newcomers. Then, as European and American awakened to the "natural world," there was another land speculation fever focused on the rivers, the lakes, and the wetlands. Many of our people were still living completely from the regenerating waters and lands until the 1940s and 1950s. This land rush, combined with a blizzard of "natural resource" exploitation of the "game," culminated in yet another displacement of our people. This occurred just when you were beginning to pass federal, state, and local laws that protect the ecology, Native American religion, burial grounds, and ancient sites.

Much of the talk today centers on means to preserve and conserve the "best" of these "wilderness" areas, while also assuring that "private landowners" are protected from undue government regulation and control over their "rights" to do what they want with "their" land. The first and last question to be asked, if not answered, is: Who owns the land? The Abenaki hold the ancient, unextinguished title here!

The History of Land Ownership and the Abenaki Homeland

Abenaki life or *Alnobaiwi*: The Abenaki way is based in this northern and southern (from Quebec perspective) forest for thousands of years, since "time unknown." Our life began with the creation and transformation of this land, passed down to countless generations in the oral tradition. For those of us from Missisquoi and other Western Abenaki places, we were made near *Bitawbagok*, Lake Champlain, by *Tabaldak*: the Creator on *aki*, the earth. We were created out of the wood of an ancient tree that still thrives here. We have always been here, kin to the ancient forests. Our life, the life of the grasses and trees, marshes, fields, and forests, is connected to us, as all human beings are joined to our mothers by the umbilical cord, as the trees and plants are rooted to the soil.

Our antiquity here is illustrated by the place names whose French or English form are still in common usage: *Missisquoi*, *Winooski*, *Lamoille*, *Connecticut*, *Ottauquechee*, *Pemigewasett*, *Winnipesaukee*, *Memphremagog*, *Umbagog*, *Merrimack*, *Contoocook*, *Saco*, *Kennebec*, *Norridgewock*, *Passumpsic*, *Mooselauke*, *Penacook*, *Amoskeag*, *Mettawee*, and many more. Behind each name are traditions linking us to the ancient life of this land.

Tabaldak, the Abenaki Creator, literally means "the Owner." When the *Awanocak* or French and *Bastoniak* or English/Americans first came here, they found it hard to understand that we feel no one can "own" the land. We share it with creation, with Creator. We fought over the use of the land with the invaders because they were destroying so much. Boundary keeping and setting territory are ancient ways necessary for most beings to live. Yet the extreme European method of total war, stealing and setting boundaries, swept into this region and made up four states, two provinces, and two countries out of our homeland in barely 200 years.

So, who "owns" the land? No person or people own the land! However, if it must be decided in the European way, we do. As in Africa, Asia, and the rest of the Americas, we Native peoples are finally coming out of the last 500 years of European expansion and genocide. *Ndakinna* is our land and has been our land for thousands of years. We are the owners with Creator because we *belong* to the land, and it will always be so. This is our only homeland!

All who live here now should be glad we have always been here. We fought and died for hundreds of years to protect this land from the hungry strangers. That struggle kept these lands and waters free of major changes until the late 1700s, long after Montreal, Albany, the cities of Massachusetts, southern New Hampshire, and the Maine coast surrounded our homeland. Now as the pollution, congestion, and terror of urban life spread north and south to the edge of this land, be glad that we still may be able to choose what happens next.

Since the time of the American Revolution, we have been dealing with the influx of settlers. During that war, Vermont, New Hampshire, Maine, and Quebec were only minimally settled by non-Abenakis. In many of the frontier settlements at Koes on the Connecticut, Pigwacket on the Saco, and Missisquoi on Lake Champlain, we Abenaki were a constant presence.

The first settlers of these regions reported extensive and often positive relations with our people. In reality, practical solutions to Abenaki continuity were worked out. That accounts for the storied Madam Campo, Captain St. Francis, Indian Joe, Molly Orcutt, Mettalac, Chief Swasson, and Old Phillip figures found widely in Vermont, New Hampshire, Maine, and Quebec histories. These family leaders were part of a living government and subsistence network that was anciently adapted to life in the northern and southern forest!

Into the nineteenth and even twentieth centuries, the Abenaki Nation has lived in linked familial networks at Missisquoi, Koes, Pigwacket, and other sections of our homeland. Our traditional subsistence life has continued, though suppressed by unjust laws and widespread abuse of the ecology. There are numerous accounts of annual Abenaki sugaring, fishing, farming, berrying, and hunting life. Local trade and sharing of subsistence skills with non-Indian people were common.

We cared for large tracts of lakeshore, swamp, woodland, and upland areas out of subsistence need, love of the land, and ancient custom. In fact, many of the state, provincial, and even federal game wardens/caretakers were and are related to us. We were and still are well suited to be intermediaries between the natural world and the human population. On the ground, we have continued to care for large sections of our homeland. Much of the independent "live and let live" way of life in the north country derives from the clear example our ancient governmental traditions provided to all.

In many ways, this century has been the hardest of times for us. Even in the old war years, including the Odanak and Missisquoi massacres of 1759–1760, the Penacook and Sokoki/Turners Falls massacres of 1675–1676, and the Pigwacket and Norridgewock massacres of 1724–1726, much of our homeland remained with us. However, the conservation/eugenics/xenophobic times from the 1880s to the 1950s have been, in some ways, more destructive than any previous period in history. Increasing control of birth, death, schooling, border travel, fishing, hunting, gathering, and remote land use have combined with a growing intolerance of any human differences to oppress our people as never before. The growth of the welfare state and the elitist conservation movement created a multipronged assault on Abenaki traditional life.

In the 1920s, 1930s, and 1940s, America and the states made a direct assault on the Abenaki similar to the war on Jews, Gypsies, and others in Europe. This movement had devastating and sometimes deadly impacts on several interconnected Abenaki families. In the process, the old network of Abenaki land, river, island, and lake caretakers has been largely replaced with new state and federal bureaucracies wholly ignorant of ancient local agreements to care for the land and waters. Many kinds of borders and walls have been erected. The extensive birth, child care, and healing networks have been severely tested. This ancient health care system was still used at the beginning of this century by most north country people regardless of background! The irony of the recent birthing room and home birth movement, as well as the most recent science-based statements that affirm the necessity to preserve both ancient Native traditions and the ecology from which we derive, is not lost on the world's Native Nations long since forced to conform to the Euro-American view of reality.

Present Situation: A Warning

First and last, you are on our land, our *only* homeland. We have been and will always be here. If you do not respect and care for our homeland and come to peace with us here, all of our children and our children's children will be left a polluted legacy of conflict and destruction in which to survive. If we must fight over the ownership of this land and the protection of the waters and creation, we will as we have in the past.

Yet understand that our land and life have already changed the invader. We showed you the way to live, the annual cycle including sugaring, farming, gathering, and hunting. We shared our life with refugees of all countries who have come here since ancient times. Many of the immigrant children married the children of this land to become rooted here. Many newcomers have come to respect this land even as we do.

Most recently, with the creation of public parks and lands, with the struggles over pollution of the land, air, and water, overdevelopment and many related issues, there are some voices among you that are sounding clearer. You are finally beginning to understand our way. You will have to come to understand sooner or later that only creation "owns" the land.

Why, you might ask, is this necessary? One answer that you are just beginning to hear is very simple: Every breath, all food, every liquid, every part of your body, your home, your physical life comes from *aki*, the earth. You are part of the earth, we are all part of the earth. We belong to creation, to the earth.

When all the trees were cut 200 years ago the rivers flooded, the soils eroded, the land cried out, and the living got hard. When the waters were filled with sewage and waste, the rivers and lakes began to smell like a dying person. And now you are just beginning to realize that no matter how rich or protected you are, the diseases made by humans are growing in you and your children. Learning to live with the earth, with the ancient knowledge of all beings who have made this home for countless generations, is a simple matter of survival. Your survival is at stake here, whether you know it or not.

Ancient Views/Future Views: A Warning and an Invitation

Beyond survival, there is a need to make peace in this war-ravaged land. It is not just the old "Indian" wars that remain unresolved in this region and the rest of the Americas. The wars on the beaver, the bear, wolf and catamount, the deer, moose, woodchuck, passenger pigeon, crow, jay, and mouse are just winding down. Insect, fish, and plant wars are still being waged all over this land. From the gooseberry wars of the 1930s to the lamprey, mosquito, and rabies wars of this decade and the coming zebra mussel war, there is still a "siege" or "fortress" mentality that seeks protection, vengeance, and conquest at the expense of the natural world. One might think at the rate the wetlands continue to be filled up that you are still at war with the evil "morass" of European folklore! Some have even been at war with our ancestors, no matter how ancient! They have dug up thousands of our ancestors' remains. These *Kitsia*/Ancient Ones, whom we know to be intimately rooted to the balance of life, have been herded into museums, washed down rivers flooded by dams, and looted for their sacred grave goods! *These wars, this way of doing things, will have to stop for this land, for you, and for all of us to come to peace on this land!*

We notice that you are still thinking "north and south" as you have since the colonial days. Your forest protection efforts run north/south in the Adirondacks, the mountains of Vermont, New Hampshire, and Maine. The new "green" corridors on the lakes and rivers have a similar form. Consider that it is equally important to think east and west to maintain corridors of life between all the mountains and river systems as we have known since ancient times. *Bitambagok*, Lake Champlain, *The Lake Between*, is not only a north/south running river or channel to go up to Albany and New York or down to Montreal and Quebec! It joins the mountains of the west and the east, as well as ways that run northwest to southeast and northeast to southwest! Bears and many other beings used to live on the islands there, swim from one side of the lake to the other, and even migrate in large numbers that way. If we want to end the wars on these beings we can learn to quiet down our human way of blocking the ancient travel routes so the rest of creation remains free to travel.

The moose have just begun to return to our homeland in large numbers. They follow the marshlands and woods in regeneration. They are to us what the buffalo are to the Plains Native Nations. Many of the ancient names of this land honor the moose. The last large bear sanctuaries are being protected. The shad, the eagle, and the turkey are returning. The bear, the moose, and many of our other relations have shown us the ways of life, and have taught and sustained us. Their life and our life are one.

A beginning awareness has recently come to you that some of our rarest endangered relatives are worthy of protection. This is also very important. Some of our people have known and protected the rarest of our relations since the old days. Though we are distressed by the elitist heavy-handed way some agencies and private individuals/institutions have lately taken over major marshes, mountains, and other places crucial for this bird, that turtle, this rare plant, that small fish, we understand that this last-moment rush to save these beings is probably necessary and well intentioned. You must, however, move as clearly into the schools and places where regular people live and explain the necessity of all this much better. There is a growing war between the freedom of people to use "their own land" and those who want to preserve the "vanishing" natural life that threatens the best efforts of both sides. We see the struggle that rages between those who use things like state and national parks and Vermont laws like Act 200 and 250 versus those who believe that private individual landowners are "king" and that the "free market economy" should be allowed to determine the future of the land. Both ways have a positive intent based in Euro-American struggles for "freedom and justice" and "life, liberty and the pursuit of happiness" for all human beings.

People, humans, must ultimately *choose* to do the right thing. No government has ever succeeded in channeling the human way. And yet, it is each individual who flushes waste in the waters, grabs the special lands for exclusive private use and posts the land, takes the "me first" approach that causes some of the most destructive patterns of human life.

Our ancient government is a "true democracy" as noted by some historians in the eighteenth century. Our political freedom within families and within the Nation were copied in some ways when you created the New England town meeting government and many other aspects of the American/Canadian democratic system of government. People yearn for freedom.

The people in the northern/southern forests of New England, New York, and Quebec must choose to care for the land. The land, the waters, and the ancients will teach even the most confused humans how to die as well as how to live. There is still room for everyone on the land. We must all choose to live together on this earth. The teaching of the ways to live balanced lives will inevitably become popular if your children's children are to survive and be at peace in *Ndakinna*. Ultimately, there is no other way than the choice to live in harmony with the earth. Listen to the land, our ancient stories, and learn.

We had best join together to protect all creation, no matter how rare or common. There are many medicines, many ways of medicine, healing, about which the Western world has only the vaguest notion. These ancient ways live in us, in our homeland. They must be protected even if you do not understand why or how they "work." Our burial grounds, subsistence grounds, sacred places, and gathering, hunting, fishing, and farming grounds are all of great importance in understanding the past, living in balance in the present, and finding the way to harmony, peace, and clarity in the future. Perhaps our ancient ways of seeing, understanding, and keeping all life in balance will be understood in an open-handed way as the future unfolds.

Of course, many aggressive companies, starting with Sir Walter Raleigh's failed tobacco enterprise and the more successful Hudson Bay Company, have made their way to the Native Nations to ask for, if not steal, specific "cures" for "diseases," foods for subsistence, etc. If the *Nahuatel*/Aztec Nation had just a fraction of the royalties for *choclatl* or chocolate, it might put some of their land issues with the Mexican government on a different footing! So you must understand our caution at sharing knowledge like the Quechua quinine cure for malaria's chills, the corn and potato food revolution, and other Native gifts that dramatically fueled the worldwide crush of conquest in the last 500 years.

The Invitation

Ask us and be patient. Silence or lack of response from us is neither permission nor a sign of ignorance. It never has been. We are still struggling with the rising tide of population, and the laws and forms you have brought to our homeland. Many of our traditional people feel the need to live at peace in the remnant of our world, rather than attempt to convert or teach those just arriving "off the boat."

The Future of the Northern/Southern Forests

Many Native Nations, including our Passamaquoddy and Penobscot cousins of Maine, own and manage their own woodlands. They have negotiated access issues with the various state and federal agencies on public lands as needed. Many of our people live from the land, from the trees, as other north country residents do. This region is trying to heal to become as it was hundreds of years ago. Yet there is so much more that must be done. It is barely a beginning. Already the hunger for the "good life" grows and threatens to engulf the region in another wave of conquest and exploitation. Thousands of acres of forest land are being clearcut for profit in preparation for the largest assault on the woodlands since the war days of the eighteenth century!

How can this be avoided? The simplest way is to understand that you are on someone else's land! Think of this as *Ndakinna*, "Our Land" or "Creator's Land." All the lands that are as yet "undeveloped" we all should choose to leave for regeneration of the ancient forests and life. Choose a way that means no more mountain developments, no new dams, no lake or river floodplain development, none. Choose to stop purposely spreading foreign species, from the Mediterranean grasses to the zebra mussel, purple loosestrife, English sparrow and all! And choose to stop the toxic wars on any beings in creation to suit the few. Stop trying to "fix" everything with another human-created "cure" without understanding life's awesomely intricate balance. Start cleaning up your mess!

And we Abenaki must have the right, explicitly outlined if necessary, to live our ancient way in our homeland. Our elders are still maintaining the legacy that has sustained us for countless generations. We are the keepers of the earth and waters of *Ndakinna*. The millions of our ancestors and the vast time in which we have lived here provide a root that serves all creation's survival in a good way. There has always been room on this land for us. Open your hearts and clear your minds. We are still here in the northern/southern forests of Vermont, New Hampshire, Maine, New York, and Quebec!

3

The Northern Forest: Problems, Politics, and Alternatives

Christopher McGrory Klyza

The future of the Northern Forest, an area of nearly 26 million acres in northern Maine, New Hampshire, Vermont, and New York, surfaced as a major issue in the mid-1980s. This area is valuable ecologically, is crucial open space in the densely populated Northeast, and has traditionally been part of an important timber-based economy. The catalyst for the emergence of this policy debate was the fear that the forest was being fragmented and converted to nontimber uses, that forest products jobs might disappear, and that this land, traditionally open for recreation, might become closed to the public.

There is limited federal land ownership in these four states: the Green Mountain National Forest in Vermont (345,000 acres), the White Mountain National Forest in New Hampshire and Maine (over 750,000 acres), and Acadia National Park in Maine (35,000 acres) are the main federal holdings. Of these holdings, only parts of the White Mountain National Forest are actually in the boundaries of the Northern Forest study area (as defined by the Forest Service and the Governors' Task Force on Northern Forest Lands). There is substantial state ownership, though, with approximately 2.6 million acres owned by New York in the Adirondack Park and over 200,000 acres in Maine's Baxter State Park. In the entire Northern Forest study region, 16% of the land is in public ownership.[1]

The bulk of the private land is held by industry (nearly 10 million acres) and large nonindustrial owners (3.6 million acres). The majority of the Northern Forest, 15 million acres, is in Maine, and it is here that large industrial and private ownership is centered. As indicated in Table 3.1, the

11 companies that own more than 400,000 acres in Maine control a total of 9.3 million acres.[2] In New England, the vast bulk of this forest land has traditionally been open for recreation. For example, of the 10.6 million acres owned by corporate and individual members of the Maine Forest Products Council, 10.3 million acres are open to public recreation.

Although the Northern Forest policy debate is generally about land conversion, three related concerns have arisen. First, the forest products industry wants to see changes in the tax code that are more favorable to timber ownership. This, the industry argues, will help prevent land conversion and protect the working forest. Environmentalists have focused on two concerns. First, almost all environmental groups have agreed that increased land protection is necessary in the Northern Forest, through easements or, more preferably, through land acquisition. And second, some environmental groups have focused on particular forest practices on private lands, practices that they argue are environmentally destructive. In addition to the controversy stimulated by these concerns, the scope of the policy debate has led to further conflict. Local residents and the forest products industry argue for local control, while others, including many environmental groups, insist that the Northern Forest is a national issue. To support their position, they point out that the Northern Forest Lands Study received more letters from California than from Vermont.[3] The controversy is further reflected in the dispute over the mission statement of the Northern Forest Lands Council, which raged through late 1992.

This chapter is an introduction to the Northern Forest policy debate. In the sections that follow, I will examine how the Northern Forest problems made it onto the political agenda, discuss the political context in which these policy concerns arose, examine the alternatives suggested to deal with the Northern Forest problems, provide a roster of the major actors in this political process, and discuss how certain issues, political currents, and alternatives may connect to lead to the adoption of a particular policy.[4]

Problems in the Northern Forest

Before examining the specific problems in the Northern Forest, it is important to examine certain larger trends affecting the area. A number of important trends were, and are, underway nationally in the forest products industry, each of which affects the Northern Forest. First, the real price for lumber increased less than 10% between 1950 and 1986. As Roger Sedjo writes, "The relative long-term stability of lumber price behavior for the 1950–1990 period suggests the possibility of a fundamental change in the long-term balance in the supply and demand for industrial wood in

TABLE 3.1.
The Largest Landowners in the Northern Forest of Maine as of July 1992

Name	Acres	Headquarters
Bowater	2,088,432	Darien, Conn.
Seven Islands Land Co.	1,011,000	Bangor, Maine
International Paper	980,891	Purchase, N.Y.
Prentiss & Carlisle Management Co.	970,000	Bangor, Maine
S. D. Warren (Scott Paper)	930,000	Philadelphia, Pa.
Champion International	730,000	Stamford, Conn.
Boise Cascade	670,000	Boise, Idaho
Irving Pulp and Paper	561,000	Saint John, New Brunswick, Canada
Diamond Occidental/James River	526,000	Richmond, Va.
Georgia-Pacific	488,035	Atlanta, Ga.
J. M. Huber	405,000	Edison, N.J.

the United States." One of the main reasons for this relative price stability is that nationally "by 1987 the total inventory of growing stock on the nation's timberland was 27 percent larger than it had been in 1952. . . . As the stocks of the nation's forest resources have increased, the pressures of real rising prices have abated." There has been a significant increase in prices in the spring of 1993, though. This is due to the combination of increased demand caused by Hurricane Andrew and a recovering housing market, and decreased supply due to timber cutting restrictions in the Pacific northwest resulting from habitat protection of the spotted owl under the Endangered Species Act.[5]

A second important trend is the increased import of wood products to the United States. By the early 1980s, imports from Canada constituted 30% of the lumber consumption in the United States. Third, major technological innovations altered the forest products industry. For example, species that were once useless to industry are being used in new products (e.g., aspen for waferboard), wood-saving devices are leading to increased efficiency in the use of harvested wood, and increased mechanization of the forest products industry has led to a reduction in employment in many parts of the country, even as harvests increased.[6]

Finally, a major shift in the geographic focus of the forest products industry is taking place. Over the last 20 years, the industry has begun to shift from the Northwest to the South, where there has been a tremendous increase in plantation forestry. In 1950, 500,000 acres of trees were being planted. In 1987, 3 million acres per year were being planted, primarily in the South (2 million acres). The South became the prime region for plantation forestry because of biological conditions, shorter harvest rotation (and hence better financial returns), flat lands (lower cost management), well-developed infrastructure, inexpensive land, good species, and good access to domestic and international markets.[7]

In New England, a major trend over the last century is the return of the forest throughout the region. In much of the area, the forest had been cleared for agriculture, but as agriculture declined in the region the forest returned. From the mid-1880s to 1980, the percentage of area forested increased from 74% to 90% in Maine, from 50% to 86% in New Hampshire, and from 35% to 76% in Vermont.[8]

The catalyst that played the chief role in identifying the Northern Forest as a problem and helping to get it on the agenda was the 1982 takeover of Diamond International by Sir James Goldsmith, a British financier. Although this event focused specifically on the Northern Forest, it was also part of larger national trends: the surge in takeovers in the 1980s and, more specifically, the special attention focused on forest products industries for takeover during this period. Goldsmith took over Diamond, which owned just under 1 million acres in the four Northern Forest states, in December 1982. Within 8 months he had sold off nearly all of the divisions of the company, earning nearly 90% of the $660 million he paid. Most importantly, he was still left with all of the land (1.7 million acres), worth an estimated $723 million. The arithmetic demonstrates why this takeover made good financial sense, and it also demonstrated why many other forest products companies were vulnerable: undervalued assets due to high values in unharvested timber and the development potential of land. Goldsmith was now eager to sell the lands and move on.[9]

In 1987, Goldsmith sold what remained of Diamond International to Cie Generale Electricite (CGE), a French communications and holding company that he once controlled. The state of New Hampshire and The Nature Conservancy had talks with Goldsmith/CGE about acquiring the 67,000 acres of Diamond Land in the state, but they could not agree on a price. CGE asked for $19 million for the 90,000 acres in Vermont and New Hampshire ($212 per acre), then agreed to sell just the 67,000 acres in New Hampshire for over $14 million (still $212 per acre). New Hampshire backed away from the offer.[10]

In May 1988, Rancourt Associates, a development group based in New Hampshire, bought all 90,000 acres, planning to subdivide at least some of the land for vacation homes. It was this purchase that really began to focus environmental, governmental, and public attention on the Northern Forest.

Just 2 months later, the state of New Hampshire bought over 45,000 acres from Rancourt, including the entire Nash Stream watershed, with a mix of state and federal money. The state's delay in purchasing the land, however, was costly. They paid $12.75 million for 46,700 acres ($282.50 per acre), giving Rancourt a $3.1 million profit for owning the lands for a few months. Rancourt also retained gravel mining rights in Nash Stream. The state supplied $7.5 million, and would manage timber harvesting and recreation. The federal government, for its $5.25 million, received the development rights to the land. Many in Congress were upset with the outcome of the Nash Stream purchase because they thought the federal money was to help in the expansion of the White Mountain National Forest, not state land acquisition. Bruce Vento, chair of the House Interior Committee's Subcommittee on National Parks and Public Lands, said, "I think it's a sham. It's going to cast a real shadow over any money going into New England." The purchase also demonstrated conflict within the environmental community. National groups such as the Sierra Club and the Wilderness Society were not thrilled with the outcome, while the Society for the Protection of New Hampshire Forests played a key role in the process.[11]

Meanwhile, there was action on the former Diamond Lands in other states. Of the 23,000 acres for sale in Essex County, Vermont, The Nature Conservancy loaned the state $1.9 million to buy 7,600 acres. In Maine, where nearly 800,000 acres were for sale, the land was sold in pieces: 9,400 acres to The Nature Conservancy in June 1988, 230,000 acres to the Fraser Paper Company in August 1988, 40,000 acres to The Nature Conservancy in October 1990, and 52,000 acres over the years in a variety of smaller sales. A partnership of James River and Diamond Occidental controls the remaining lands in Maine.[12]

And in New York, the state also ended up having to buy the land from a developer. Lassiter Properties of Georgia bought the 96,000 acres available in New York state for $17 million in the winter of 1988. The state then purchased 15,000 acres of land and development rights on 40,000 acres in the Adirondack Park from Lassiter for $10.4 million. This purchase was part of a larger set of concerns with development in the Adirondacks. In response to these and other concerns, Governor Mario Cuomo created the Commission on the Adirondacks in the Twenty-First Century to make recommendations to protect the park into the future.[13]

These Diamond land sales led environmental groups to sound the alarm that the Northern Forest was about to face strong development pressures. The Wilderness Society warned that in the 1980s "a growing boom in second-home and resort development not only began to limit public access but threatened to destroy the wild character of some of the finest land left in the region." The Sierra Club claimed that "millions of acres of private timberland are coming on the market. This turnover provides a once-in-a-lifetime opportunity to establish new public lands in a heavily populated region where only 3 percent of the land is federally owned. . . . Quick action is essential here, before the prime lands on the market are converted to private subdivisions." [14]

This fear of second-home development in the Northern Forest was exacerbated by the economic boom that was occurring in New England and the northeast at the time. Many people in the area saw their incomes rise tremendously, and they had money to invest in second homes. The real estate prices reflected this increased demand. For example, a 40-acre lot on Moosehead Lake in Maine that sold for $50,000 in 1986 sold for $250,000 in 1988.[15] Many were quick to point out the large amount of land that was on the market. In 1990, it was reported that International Paper had 237,000 acres on the market, and Boise Cascade, Georgia-Pacific, and the Penobscot Indian Nation had a total of 223,000 acres for sale.[16]

Other takeovers and mergers were also occurring. The 1984 merger of St. Regis and Champion International created the largest paper company in the United States. In 1985, Goldsmith acquired Crown Zellerbach in a hostile takeover, and James River purchased part of the assets of Crown Zellerbach. The previous year, Goldsmith was involved in unsuccessful bids to take over the Continental Group and St. Regis. In February 1990, Georgia-Pacific completed its takeover of Great Northern, which owned over 2 million acres in Maine. Less than two years later, in October 1991, Georgia-Pacific sold 80% of the Great Northern land to Bowater, with an option for the remaining 20% for $322 million.[17]

Many in the region, including environmentalists, were alarmed by these events, arguing that takeovers increase the focus on maximizing returns and can lead to overharvesting, selling off land, charging fees for recreation, and using intensive management practices. Some groups argued that government should seek to block these takeovers through the public purchase of land, changes in tax policies, land-use planning, the purchase of conservation easements, and the regulation of junk bonds.[18]

Although the catalyst for the Northern Forest was the sale of forest products lands to developers and the fear that the large-scale private ownership of this land by forest products companies might be significantly reduced, environmental groups soon added another issue to the agenda: forest practices. Clearcutting on a large scale started in Maine in the late 1970s, along with the herbicide use that often accompanies it. In 1988, George Wuerthner wrote that "some observers claim timber companies are high grading their lands in a last-ditch effort to wrench some profit before abandoning them and the Northeast timber industry with them." The Sierra Club argued that current forest practices were a problem because of unsustainable harvests, construction of new roads (especially due to the end of river log floats in Maine in 1976), clearcutting, the use of herbicides and pesticides, high grading the forest (cutting only the best trees and leaving those remaining), and the loss of jobs. Relatedly, environmentalists pointed to other threats to biodiversity in the Northern Forest: air pollution, development, some recreation use, some hunting, fishing, and trapping, and a 1989 proposal by the Maine National Guard to establish a 720,000-acre Deepwoods Training Area on lands obtained from Champion International.[19]

The forest products industry also saw a set of problems in the Northern Forest. With the increasing globalization of capital, these problems focused on their difficulty in competing with other regions and other nations in an increasingly competitive market. The industry had three major complaints: first, that it was being overregulated; second, that changes in federal tax law in 1986 eliminated the capital gains policy of taxing most income from timber sales at a lower rate; and third, that high estate taxes on large family ownerships have made it hard for these large private forest landowners in the Northern Forest to continue in the forest management business. Another complaint, one that varied from locality to locality, was high property taxes.[20]

In summary, a series of trends combined with a crucial event to help place the Northern Forest on the political agenda. The trends were the changing nature of the forest products industry, the increase of mergers and acquisitions, an increase in the amount of land for sale in the Northern Forest, a change in forest practices in the Northern Forest, and an increase in development in the area. Some of these trends may not have differed much from previous years, but the perception was that changes were occurring. The catalyst was the takeover of Diamond International by Sir James Goldsmith and the subsequent sale of nearly 1 million acres in the Northern Forest. The sale of these lands to developers in New York, Vermont, and New Hampshire helped to push this issue onto the national agenda. And finally, government representatives in the area began to hear from constituents, environmental groups, the forest products industry, residents, and visitors that something needed to be done on the Northern Forest.

The Political Setting

The political setting or context in which the Northern Forest arose on the agenda included six major aspects: the importance of the environment in New England, the power of the region in Congress, the relationship of the states in the region to the federal government, the political power of the forest products industry in Maine, the rise of the property rights movement, and the recession that hit the area in the late 1980s.

In public opinion polls, people surveyed in the New England region have consistently ranked among the most supportive of environmental concerns. For example, a 1980 poll conducted for the Council on Environmental Quality indicated that New England ranked third in support for the environmental movement (67% were sympathetic) and ranked first in lack of opposition to the movement (0% were found unsympathetic). Many of the people who live in the area have strong ties to the environment, whether it be the ocean, the mountains, or the forest. Indeed, many of the people who move to the region are environmental supporters drawn to this landscape. The increasing commitment to government protection of the environment in the area is underscored by Maine's passage of a $35 million bond for land acquisition in November 1987 by a 65% to 35% margin. Although polls in Maine suggest that most prefer a working forest rather than large-scale federal ownership or wilderness area, the passage of the Bond Act is a sign of further commitment to environmental protection there. Environmental groups are also quick to point to a 1988 poll of residents in the Northeast Kingdom of Vermont by Sterling College that found 67% of those surveyed in favor of a Northeast Kingdom National Park. A 1991 Forest Service survey of residents of northern New Hampshire and Vermont found strong support for public land acquisition for a variety of uses: 85% supported public land acquisition to protect wilderness, 81% to maintain wildlife habitat, 80% to maintain recreational opportunities, and 72% to assure timber supply.[21]

The delegations of New York, Vermont, New Hampshire, and Maine are powerful ones in Congress. In the Senate, Patrick Leahy of Vermont chairs the Agriculture Committee and George Mitchell of Maine is the Majority Leader (though he is not seeking re-election in 1994). In the House, New York has the second largest delegation. It was through Leahy's Agriculture Committee, which deals with forestry issues as well, that most of the Northern Forest–related legislation moved. It was Leahy and Warren Rudman of New Hampshire who intitiated the Northern Forest Lands Study in Congress and helped gain the $250,000 appropriation for it in September 1988. Later, the 1990 Farm Bill authorized the continuation of the Northern Forest Lands Study, appropriated $1.075 million to the Northern Forest Lands Council (NFLC), and authorized the Forest Legacy Program (discussed later). Alternatives offered by environmentalists calling for large-scale purchases of public lands in the Northern Forest cite the political power of the entire Northeast region, arguing that the Northern Forest is a concern for the entire region, not just New England and New York. Viewed from this perspective, the size and power of the area's delegation increase further.

The New England states, especially the northern three, have a strong sense of independence, and this often manifests itself in a desire to limit the role of the federal government in the region. On the Northern Forest, this is revealed in the frequent statements that the federal government should not get involved, and that federal land ownership, whether through national forests or national parks, is not needed or wanted. A political reflection of this is the creation of both the Governors' Task Force on Northern Forest Lands and the Northern Forest Lands Study. Rather than a single study, the four states wanted a study group independent of the federal government to examine the issues. The NFLC, created after the Northern Forest Lands Study and the Governors' Task Force, has 16 of its 17 members appointed by the four governors of the involved states.[22]

The forest products industry is the dominant industry in Maine (40% of state manufacturing output and 10% of the active workforce are tied directly to Maine forests), the predominant landowner in the state, and the major source of jobs in much of the state. When the industry speaks, the state listens. And since most of the Northern Forest is in Maine, the state has a very strong voice within the NFLC. Therefore, the Maine forest products industry has the potential to dominate the NFLC process. However, since nearly 60% of the Northern Forest is in Maine, this is not surprising.[23]

Perhaps the most important aspect of the political setting has been the rise of the property rights movement in the region. The issue of managing public lands and the regulation of private lands had risen to the top of the national environmental agenda by the late 1970s. This was first seen in the Sagebrush Rebellion, which focused on the transfer of federal public lands to the states. This rebellion was defused in the early 1980s, and the privatization initiative rose in its wake. Privatization advocates argued that most public lands should be transferred to the private sector, where they could be managed most efficiently for any purpose, be it timber or wilderness. This initiative faded by the mid-1980s, and yet another related movement arose soon after: wise use. Like its predecessors, wise use focuses on private access to public lands and the resources on those lands, and is centered in the west. Also like its predecessors, it focuses on the importance of property rights. Although the Northern Forest features limited public lands, these movements and initiatives helped to stoke the politics of the debate. Property rights became a key to the debate for many conservatives in their battles with environmentalists. In the Northern Forest, this focused on the fear of government regulation of private property and on the government forcing people to sell their lands.[24]

Property rights advocates have been a loud voice in the Northern Forest debate. The release of the report by the state commission on the future of the Adirondacks in the spring of 1990 set off a storm of protest among real estate interests and property rights groups. Supporters of these interests have operated both within the system and outside it. In the Adirondacks, property rights advocates have attacked members of Earth First!, allegedly shot at cars belonging to the Adirondack Park Agency, and threatened members of the NFLC at a 1991 meeting in Ray Brook, New York. At hearings on the NFLC in the summer of 1991, property rights advocates were on the offensive in Bangor, Maine, and Lyndonville, Vermont. Among the groups that have been most active are the Adirondack Blue Line Confederation, Adirondack Fairness Coalition, Adirondack Solidarity Alliance, Citizens for Property Rights (Vermont), John Birch Society, and Maine Conservation Rights Institute.[25]

Property rights advocates have demonstrated their power most clearly in two instances. First, they helped prevent the passage of legislation to authorize and direct the NFLC. In the summer of 1991, all eight senators in the study area had agreed to draft legislation for the Council. In the House, though, Representative Gerald Solomon of New York led the opposition, largely in response to property rights advocates who feared that the NFLC would infringe on their property rights. In March 1992, Senators Mitchell and William Cohen of Maine withdrew their support of legislation to authorize the NFLC in response to opposition by property rights groups, the lack of support of Maine Governor John McKernan (also responding to property rights advocates), and the general lack of consensus within the state. In the end, only Senator Leahy continued to support the legislation. Nevertheless, the NFLC held its first meeting in June 1991 and has continued its work on Northern Forest issues.[26]

Second, these interests helped to stymie action in response to the recommendations issued by the Commission on the Adirondacks in the Twenty-First Century. These recommendations included the purchase of over 650,000 more acres of land by the state, a more powerful Adirondack Park Agency to regulate private land use, and a 1-year moratorium on development of private lands throughout much of the park. In response to strong opposition from property rights groups, among others, Governor Cuomo distanced himself from these recommendations. In addition, the $800 million Environmental Quality Bond in New York, half of which was to go toward land acquisition, was defeated in November 1990. This loss was due to both the efforts of property right supporters and other opponents in the Adirondacks and the recession.[27]

The recession of the late 1980s and early 1990s is the final aspect of the political setting. The recession made it difficult for government at both the state and federal level to make funds available for the purchase of land. This was demonstrated in the 1990 defeat of the environmental bond in New York, mentioned earlier. In addition, some of the development pressures of the mid-1980s seemed to dissipate. The New England and northeast economies were hard hit, and second home development became less of a threat. Indeed, the two developers who bought the Diamond Lands, Rancourt and Lassiter, both went bankrupt. Although it is likely that the region will soon be in a recovery, the recession certainly influenced the political context.[28]

Major Alternatives

Different actors in the political process have developed and supported different alternatives to achieve their goals. Environmentalists have focused on land acquisition and protection and, to a lesser degree, forest practices. The forest products industry has primarily focused on changes in taxation. The Northern Forest Lands Study, the Governors' Task Force, and the NFLC have all explored a broad array of alternatives for the Northern Forest.

Beginning in the late 1980s, environmental groups started to offer alternatives for protecting the Northern Forest. A 1988 article in *Wilderness* described four proposals in northern New England that centered on the purchase of public lands. These included expanding the White Mountain National Forest in New Hampshire by purchasing the Mahoosuc Range and the Nash Stream drainage; adding the Taconic Range to the Green Mountain National Forest in Vermont (the proclamation boundaries of the national forest were expanded in 1990 to help enable this); implementing a variety of proposals to increase public land ownership in northern Maine, ranging from a National Parks and Conservation Association proposal for 2 million acres in the St. John River drainage to a national park surrounding Baxter State Park (an idea that can be traced to the 1910s and 1930s) and a Maine Woods National Park, of up to 2 million acres; and finally, establishing a Northeast Kingdom National Park and Recreation Area in Vermont.[29]

The Wilderness Society unveiled its plan to establish a 2.7-million-acre Maine Woods Reserve surrounding Baxter State Park in March 1989. The reserve would rely on a mix of protection strategies: (1) federal and state purchase from willing sellers, (2) local and state regulations of land use, such as development zoning and conservation easements, and (3) incentives for private ownership protection through the tax code, such as reductions for protected areas and antispeculation taxes. The Wilderness Society opened an office in Maine to work on the plan, which has received a mixed reception in the state.[30]

In the Adirondacks, the Adirondack Council, which began as an umbrella for national environmental groups but has since evolved into a membership organization, seeks to have state ownership within the park increase from 2.6 million acres to 3 million acres, 50% of the park. Also during this period, environmental groups proposed the creation of the Bob Marshall Great Wilderness in the western section of the park. The 400,000-acre area would bring together existing state lands with newly acquired lands.[31]

A joint statement released by the Audubon Society, the National Wildlife Federation, the Sierra Club, and the Wilderness Society in October 1991 was critical of forest practices and proposed major land and conservation easement acquisition through the Forest Legacy Program (discussed later) and a new Weeks Act (the 1911 law that authorized national forests in the east). There would be publicly owned wildland preserves with buffer zones, featuring a mix of private and public ownership where regulated use would occur. Natural corridors would connect the core areas. This program is necessary because "the social compact has been broken—overridden by the response of the forest products industry to global economic realities." The funding for such purchases was to come from the Land and Water Conservation Fund, which receives up to $900 million annually, mainly from offshore oil and gas revenues. These groups argued that with the political power of the northeastern states in Congress, and the need for public land in the area since there is so little, funding could be made available.[32]

The Northern Forest Alliance, an association of 15 conservation and environmental groups formed in 1990, offered a more specific proposal on December 31, 1991. The Alliance identified nine jewels to be protected, eight of which were on the market and one of which could come on the market at any time. This "New Year's Resolution for the Northern Forest" would protect approximately 400,000 acres for an estimated $50 million. In March 1992, the Northern Forest Alliance called on Congress to appropriate $50 million for the Forest Legacy Program in the Northern Forest for Fiscal Year 1993. It should be noted that the Northern Forest Alliance has had difficulty agreeing on common positions since it represents a broad spectrum of environmental groups. This struggle has also limited the effectiveness of the alliance in the political realm.[33]

The most far-reaching environmental alternatives have been offered by Preserve Appalachian Wilderness (PAW), a grassroots group initially affiliated with Earth First! and now connected to Earth Island Institute. In its response to the Northern Forest Lands Study (NFLS), "The Working Forest Is Not Working: A Critique of the Northern Forest Lands Study," PAW finds the NFLS to be pro-timber, without enough focus on biodiversity and forest health. The PAW report focuses on problems with forest practices, and concludes by recommending the spending of $300 million per year for the next decade to create a Northern Appalachian Evolutionary Preserve of 10 million acres. In the preserve, wilderness cores would be connected by wild corridors and surrounded by buffer zones. The area included in the preserve would be expanded to include the Green Mountain National Forest, the White Mountain National Forest, the Gaspé Peninsula, New Brunswick, the Taconics, the Catskills, and the Berkshires. The recovery of extirpated species such as wolf, lynx, cougar, and wolverine would be of central importance in the Preserve.[34]

Although not endorsing acquisition programs of such scope, and not endorsing the purchase of specific areas, the NFLS and the Governors' Task Force on Northern Forest Lands (GTF) also recommended the use of the purchase and easement strategies. The NFLS identified a host of conservation easement and purchase strategies, including conservation easements, land acquisition with management by the state or federal government (in national forests—existing or new, new national parks, or new wildlife refuges), rolling leases (a kind of temporary conservation easement), and the acquisition by government agencies of the right of first refusal for certain lands. The GTF recommended increased state land acquisition by fee and conservation easements, and grants to the states from the Land and Water Conservation Fund of $25 million a year for 4 years (a total of $100 million) that the states would use to purchase conservation easements or land. One new program has already been approved based on these proposals. The Forest Legacy Program, established in 1990 as part of the Farm Bill, is designed to protect environmentally important forest lands threatened by conversion to nonforest uses, primarily by purchasing conservation easements from willing sellers. This program, administered by the Forest Service and the individual states, received appropriations of $5 million in Fiscal Year 1992.[35]

Many of the proposals to alter the tax system to make forest ownership more attractive for those involved in forest industries were discussed in the NFLS and by the GTF, and have been further studied by the NFLC. Strategies discussed in the NFLS include favorable capital gains tax treatment for timber from managed timber lands, estate taxes based on current use if agreements are made to keep the land undeveloped for the next generation, tax credits for investment in forest land or resource-based industry, deduction of timber management costs if landowners agree not to develop land, and current use valuation of land for property taxes. Among the strategies recommended by the GTF are to excuse federal and state income tax on sales of land for conservation purposes, to establish tax credits for donations of land for conservation purposes, to reinstate favorable capital gains treatment for sale of timber in a 10-year holding period, to improve existing tax programs to further protect large forest tracts, and, relatedly, to limit landowner liability.[36]

The focus of the NFLC on these alternatives is clear; the Council has subcommittees on Property Taxes, State and Federal Taxes, and Local Forest Based Economy, and related economic-oriented subcommittees on Land Conversion and Recreation and Tourism. It has also produced the following studies: "Federal Taxation and the Northern Forest Lands," "Property Taxes and the Economics of Timberland Management in the Northern Forest Lands Region," "Forum on Forest Based Economic Development in the Northern Forest," "Forum on National and International Influences on Land Conversion," and "Forum on Land Sales of Coburn Lands Trust and Former Diamond International Corporation." An April 1992 letter from the NFLC to the congressional delegation of the four states further underscores the importance of this approach. The letter focuses on "major areas of current federal taxation policy that provide significant disincentives to the maintenance and the stability of the valuable open space areas of the Northern Forest." It lists six concerns that the members of Congress should examine: the investment classification of forest ownership and material participation, the amortization of reforestation expenditures and the accompanying tax credit, the differential treatment of income from the sale of timber, the allowance of the deduction of normal timber management costs as expenses, conservation buyer market advantages, and estate tax policies.[37]

There are two other major alternatives. The first is the use of planning and land-use controls. For example, in Maine the Land Use Regulation Commission (which regulates land use in unincorporated parts of the state) could use Natural Character Management Zoning, a type of current use zoning that allows forestry, agriculture, and primitive recreation, but no development. This approach is available, but is not used due to landowner opposition.[38] The second major alternative is one that can be incorporated with a number of alternatives previously discussed: the creation of green line areas. Programs used within the green line area to help maintain the Northern Forest could include conservation easements, land purchases, tax incentives, and current use zoning, among other approaches.[39] Other alternatives discussed in the NFLS include combining community improvement with land conservation, improving landowner liability laws, and increasing private land user education.

As discussed earlier, of the vast array of alternatives formulated and discussed, some are more likely to be seriously considered than others. Those bearing serious consideration must clear three hurdles: technical feasibility, community acceptance, and anticipated future constraints. Each of the major proposals discussed here appears to pass the first hurdle; they all rely on existing techniques that have been used elsewhere to some degree. The hurdle of community acceptance appears to loom larger for some alternatives. Indeed, the response of property rights advocates suggests significant community resistance to public acquisition and land-use regulation. Similarly, alternatives involving forest practices regulation will surely face opposition from the forest products industry. And finally, the most important future and present constraint is the fiscal situation of state and federal governments. Any proposal that will cost money, be it land purchases or altered tax burdens, faces a difficult path in a period of fiscal restraint.

The Policy Future

The purchase of Diamond International by Goldsmith in 1982, and the subsequent sale of some of these lands to developers, triggered a region-wide policy debate on the future of the Northern Forest. This debate has centered on questions of forest land conversion, forest land taxation, forest practices, public land acquisition, and private property rights. A multitude of alternatives has been developed by environmentalists, the forest products industry, property rights advocates, and others to address these concerns, and the Northern Forest Land Council is currently sorting through them. In September 1994, the Council is due to issue its recommendations to Congress and the states, and then disband.

As the NFLC nears the end of its existence, it is unclear what, if any, policy alternatives will be adopted. The issuance of these recommendations, though, will present a real opportunity to adopt new policies on the Northern Forest. The likelihood of the successful adoption of such new policies often rests with a policy entrepreneur, someone who takes a special interest in the issue and works to see a policy adopted. On the Northern Forest, Senator Leahy has been the chief policy entrepreneur. From his position as chair of the Senate Agriculture Committee, he has helped to establish the NFLS and the NFLC, to pass the Forest Legacy Program, and to get funding for the NFLC. When the authorization legislation for the NFLC failed to pass in 1991, though, Leahy seemed to lose some of his earlier enthusiasm. Whether this is actually the case will become clearer when the NFLC disbands in 1994. If Leahy is no longer the policy entrepreneur, another may arise. If significant policy on the Northern Forest is to be adopted, it would be best if the entrepreneur is powerful, and also if he or she is tied to Maine.

Regardless of the policy outcome, it is clear that the results of the Northern Forest policy process can serve as an indicator of United States ecosystem and natural resources policy entering the twenty-first century. If all parties agree to a comprehensive plan to help develop sustainable natural and human communities in the Northern Forest, this process can give us hope and serve as an example for other parts of the country. If the process ends in stalemate, with no significant policy changes, the Northern Forest will be illustrative of the gridlock in environmental policy evident elsewhere in the country. We will soon know which vision reflects reality.

4

The Northern Forest Economy

Thomas Carr

The economy of the Northern Forest region is intricately linked to the natural resource base. The timber resources are a primary input in the production of paper and allied products and a variety of wood products. Large paper mills convert the region's trees into regular white paper, newspapers, specialty papers, and cardboard boxes. Sawmills cut logs into lumber products that are used to build houses, kitchen cabinets, and furniture. In addition to these forest products, the rich soils of the valleys linking the mountain ranges support a modest agricultural sector. Numerous recreational opportunities are available in the region, including hunting, fishing, skiing, hiking, camping, canoeing, and horseback riding. These activities are the basis of a substantial regional tourism industry. Increasingly, many people are finding the special attributes of recreation opportunities, open space, and aesthetic landscapes excellent incentives to invest in a vacation home.

The objective of this chapter is to provide background on the economic factors relating to the controversy over the Northern Forests. The chapter proceeds in the following manner. In the first section, I present an overview of the Northern Forest economy. The analysis focuses on the sectors most relevant to land conversion: forest products, agriculture, tourism, and development. The economic forces causing change are examined in the second section. The analysis identifies key factors contributing to development and explores the long-run prospects of the world timber market. I conclude with some observations about the role of markets, economic theory, and policy.

The Northern Forest Economy

This section outlines the characteristics of the economy in the Northern Forest region. The following terminology will be used to differentiate the appropriate area. The *Northern Forest region* is defined as the same 25.8-million-acre area specified in the Northern Forest Lands Study.[1] The breakdown of this area by state is presented in Table 4.1. Unfortunately, economic data are not compiled within the exact borders of this region. Instead, a considerable amount of data is collected at the county level and the state level. For practical purposes, the *Northern Forest economy* is defined by combining the 28 counties in the area. This area is also called the *Northern Forest counties*. The Northern Forest counties cover 31.35 million acres, or 5.55 million acres more than the Northern Forest region. Most of the differential between these two areas arises from counties in the state of New York that are only partially in the Northern Forest region. In a parallel fashion, the entire four-state region is called the *Northern Forest states*. As presented in Table 4.1, the Northern Forest counties amount to one-half of the total land area of the Northern Forest states. In the total Northern Forest counties, over one-half of the land area is in Maine (53%), with another third in New York (37%), followed by Vermont (6%) and New Hampshire (4%).

Table 4.2 presents data on the general features of the Northern Forest economy. With an overall population of 1.8 million people, the area includes just over 500,000 jobs that generate an income flow of $11,500 per capita and a median household income of $23,600. By comparison, the Northern Forest per capita income is 80% of the statewide figure, and 83% of the national average. Similarly, the Northern Forest median household income is 82% of the statewide average, and 85% of the national level. Income certainly does not measure all factors that determine the quality of life. Table 4.3 provides additional indicators on different measures of social welfare. These factors reflect some of the inherent trade-offs associated with the costs and benefits of living in a rural region. Rural residents benefit from environmental amenities, reduced congestion, and a significantly lower crime rate. On the other hand, these residents have lower incomes, access to fewer physicians, lower educational attainment rates, and a higher percentage of the population needing governmental assistance in terms of the Supplemental Security Income Program. The condition of the economy influences all of these variables, directly or indirectly, by providing the resources to make investments in the workforce, education, and health care.

Manufacturing (25.4%) and services (28.1%) account for over 50% of total employment in the Northern Forest economy. Another 30% is attrib-

TABLE 4.1.
Distribution of Land Within the Four States of the Northern Forest

	Land Area (×1000 acres)		
	Northern Forest region	Northern Forest counties[a]	Northern Forest states
Maine	15,000	16,603	19,837
New Hampshire	1,200	1,155	5,755
Vermont	2,000	2,000	5,935
New York	7,600	11,592	30,323
Total	25,800	31,350	61,850

SOURCE: Data derived from Harper, Stephen C., Laura L. Falk, and Edward W. Rankin, 1990, *The Northern Forest Lands Study of New England and New York*, Rutland, Vt.: U.S. Department of Agriculture, Forest Service; and Rand McNally, 1992, *Commercial Atlas and Marketing Guide*, Skokie, Ill.: Rand McNally.

[a]The Northern Forest Counties in each state are: *Maine*: Aroostook, Franklin, Hancock, Oxford, Penobscot, Piscataquis, Somerset, Washington; *New Hampshire*: Coos; *Vermont*: Caledonia, Essex, Franklin, Lamoille, Orleans; *New York*: Clinton, Essex, Franklin, Fulton, Hamilton, Herkimer, Jefferson, Lewis, Oneida, Oswega, St. Lawrence, Saratoga, Warren, Washington.

TABLE 4.2.
Demographic Data for the Northern Forest Region

	Population 1990	Employment 1990	Per capita income 1990	Median household income 1990
Northern Forest counties				
Maine	465,823	139,821	11,458	22,795
New Hampshire	34,828	12,244	12,142	24,836
Vermont	118,019	31,681	10,532	22,021
New York	1,217,063	318,095	11,893	24,547
Total	1,835,733	501,841	11,534	23,606
Northern Forest states				
Maine	1,227,928	424,027	12,946	25,652
New Hampshire	1,109,252	439,636	15,913	33,753
Vermont	562,758	215,222	12,761	25,717
New York	17,990,455	7,075,441	15,890	30,330
Total	20,890,393	8,154,326	14,378	28,863
Total United States	248,709,873	117,914,000	13,952	27,912

SOURCE: Data from Rand McNally, 1992, *Commercial Atlas and Marketing Guide*, Skokie, Ill.: Rand McNally.

TABLE 4.3.
Social and Economic Indicators of Human Welfare in the Northern Forest Region

	People per square mile (1990)	Physicians per 100,000 (1985)	Education, percent with 12 years (1980)	SSI recipients per 1,000 (1986)	Crimes per 100,000 (1985)
Northern Forest counties					
Maine	18	120	67.0%	25	2,579
New Hampshire	19	151	58.7%	12	2,059
Vermont	38	102	64.1%	23	1,705
New York	67	102	64.4%	17	2,478
Total	37	109	64.7%	20	2,446
Northern Forest states					
Maine	40	164	68.7%	19	3,643
New Hampshire	123	178	72.3%	6	3,120
Vermont	61	232	71.0%	18	3,823
New York	380	286	66.3%	21	5,521
Total	216	215	69.6%	16	4,027
Total United States	70	197	66.5%	18	5,037

SOURCE: Data from Rand McNally, 1992, *Commercial Atlas and Marketing Guide*, Skokie, Ill.: Rand McNally; and U.S. Bureau of the Census, 1988, *County and City Data Book*, Washington, D.C.: Government Printing Office.

uted to retail and wholesale trade. The assortment of construction, transportation and utilities, finance, insurance, and real estate makes up about 15%. The agriculture sector employs less than 1% of the workforce. To obtain more insight about the relationship of the economy to the resource base, we must take a more in-depth look at four distinctive components: forest products, agriculture, tourism, and development.

Forest Products: Paper and Lumber Industry

For this analysis, the forest products sector is defined as the combination of the lumber and wood products industry, plus the paper and allied products industry. Timber resources are a common raw input for both types of output. In the Northern Forest region, large tract commercial forestry utilizes nearly 13 million acres of land. This represents 50% of the land area in the region and 60% of the privately owned land. Paper manufacturing firms hold about 9 million acres, or 40% of the privately held land. Wood manufacturing firms hold another million acres. Large nonindustrial landowners account for another 2 million plus acres of land. The residual private land is largely in the hands of smaller nonindustrial landowners that may or may not harvest their forests.[2]

In commercial forestry, trees are viewed as a crop that grows over long intervals of time. The forester's economic problem is finding the optimal rotation that maximizes profits for the landowner. The solution to the problem was originally derived by Martin Faustmann in 1849, and refined by P. H. Pearce and Paul Samuelson.[3] Intuitively, timber is an asset that appreciates with the growth of the forest. The optimal rotation period occurs when the capital gains from timber growth equals the opportunity cost of keeping wealth tied up in an inventory of timber. In general, theory tells us how the following items affect the optimal rotation period. Higher interest rates increase the opportunity cost of holding timber, and thereby shorten the optimal rotation period. Higher prices of timber serve to lengthen the rotation period. Different types of taxes affect the rotation period in different ways.

The simple optimal rotation theory becomes more complex when individuals have multiple objectives for holding forests. For instance, paper manufacturers reportedly hold some forests as a backup or surplus inventory of timber for their paper mills. This is effectively an insurance policy against the risk of losing normal supplies and the corresponding high costs of an idle mill.[4] In contrast, a small nonindustrial landowner may derive benefits from the standing forests in the form of hiking, hunting, or aesthetic landscapes. If the nontimber benefits are sufficiently large, the private landowner may find it optimal to never cut the standing timber.[5]

For commercial forestry, trees are grown and managed until maturity. Upon reaching the desired age, the timber is logged and then processed as lumber or paper. In the Northern Forest states, there are 889 logging establishments, 50% of which are located in Maine. This timber supports a large infrastructure of 60 sawmills, 50 paper mills, 25 pulp mills, 10 cardboard mills, 6 veneer mills, and 5 panel mills. Across the border in Canada are another 50 pulp, paper, and sawmills.[6] The spectrum of paper products includes newsprint, white paper, paperboard, corrugated boxes, and tissue paper. Lumber products are a primary input in the construction industry and the manufacturing of furniture. There is a sizable furniture manufacturing base consisting of about 100 establishments with more than 25,000 employees in New York and Vermont. In this analysis, the furniture sector was not included in the definition of the forest products industry.[7]

In 1990, total lumber output was 1,476 million board feet, 50% of which comes from Maine and 22% from Vermont. For the entire area, softwoods account for 70% of the lumber production. Vermont is unique with 82% of its output in hardwoods. Overall, Northern Forest state lumber production

TABLE 4.4.
Economic Output of the Forest Products Industry in the Northern Forest Region

	Lumber and wood	Paper and allied	Total forest products
Value added in 1990 (million $)			
Maine	531	1,797	2,327
New Hampshire	150	379	529
Vermont	135	184	318
New York	526	2,248	2,774
Total	1,342	4,608	5,948
Percentage of 1989 gross state product			
Maine	2.1%	5.5%	7.6%
New Hampshire	1.0%	1.5%	2.4%
Vermont	1.3%	1.5%	2.8%
New York	0.2%	0.5%	0.6%
Total	0.3%	0.8%	1.1%

SOURCE: Data are from the U.S. Bureau of the Census, 1990, *Annual Survey of Manufacturers*, Washington, D.C.: Government Printing Office; The Irland Group, 1992, *Economic Impact of Forest Resources in Northern New England and New York*, Northeast Forest Alliance, p. 80; and U.S. Department of Commerce, Bureau of Economic Analysis, 1992, *Business Statistics*, Washington, D.C.: Government Printing Office.

amounts to only 2.7% of the total U.S. lumber output. Paper products are really the primary product of the Northern Forests. The Northern Forest states produced 5.3 million tons in 1988, or 8.7% of the national output. Within the Northern Forest states, Maine is the largest producer with 62% of the pulp output, followed by New York at 30%, New Hampshire at 5%, and Vermont at 3%.[8]

The forest product sector generated $13.6 billion of sales across the entire four Northern Forest states in 1990. A more accurate measure of economic output is value added from the forest product sector. Value added avoids the double counting problem as sales progress up the chain of production. Table 4.4 shows that the forest product sector creates $5.9 billion in value added in the four Northern Forest states. In contrast to the high relative production shares by Maine in the production of lumber and pulp, New York has the largest share of value added with 46.6% of the total. New York's large share of value added is due to the large number of pulp and paper mills that process the raw inputs. Table 4.4 also shows forest sector value added as a percentage of each state's total output, as measured by the 1989 gross state output (GSP). The forest product sector has the largest share in Maine with 7.6%, followed by Vermont at 2.8%, New Hampshire at 2.4%, and New York at 0.6%. The New York state economy is nearly

TABLE 4.5.
Employment and Payroll in the Forest Product Sector in Northern Forest States in 1990

	Employment			
	Lumber and wood	Paper and allied	Total	Payroll (million $)
Northern Forest states				
Maine	12,531	15,906	28,437	814
New Hampshire	4,448	4,461	8,909	214
Vermont	3,301	1,908	5,209	116
New York[a]	17,854	40,625	58,479	1,284
Total	38,134	62,900	101,034	2,428
Percentage within Northern Forest states				
Maine	3.0%	3.8%	6.8%	9.9%
New Hampshire	1.0%	1.0%	2.0%	2.3%
Vermont	1.5%	0.9%	2.4%	2.8%
New York[a]	0.3%	0.6%	0.9%	0.7%
Total	0.5%	0.8%	1.2%	1.2%

SOURCE: Data are from U.S. Bureau of the Census, 1990, *County Business Patterns*, Washington, D.C.: Government Printing Office.
[a] Data are from 1989.

10 times the size of the three other states combined; thus, the small relative shares for the total four-state region reflect the heavy GSP from New York.

The forest products industry employs nearly 100,000 workers in the four Northern Forest states. As shown in Table 4.5, this represents 7.1% of all manufacturing jobs and 1.2% of the total workforce. Although New York has the largest number of forest product workers, its relative contribution is much smaller given the large size of the New York economy. In contrast, Maine's 28,000 forest product workers are 27.5% of the manufacturing workforce, and 6.7% of total Maine employment. New Hampshire and Vermont have similar relative aggregate employment levels, which account for about 10% of the manufacturing labor and 2% of total employment. The payroll figures tell a similar story about the relative importance of the forest products sector in the economy. The forest products sector payroll of $2.5 billion amounts to 6% of the manufacturing payroll and 1.2% of the total payroll. Overall, the relative shares of payroll mirror the employment shares among the states.

Employment from the forest products sector takes on even greater relative significance at the county level in the Northern Forest. Unfortunately, the precise number can not be determined from Census Bureau data. Employment and payroll figures at the county level for the paper and allied products sector are presented in broad ranges. This practice is designed to protect against the release of proprietary information for individual firms. Given the range of data provided for Maine's Northern Forest counties, the forest products sector contributes between 47% and 67% of the manufacturing jobs, and between 13% and 19% of total employment. In New York's Northern Forest counties, the corresponding ranges are between 17% and 19% of the manufacturing sector, and 4% of the total workforce.

In summary, the forest products sector provides the Northern Forest states with 100,000 jobs, $5.9 billion in value added, and $2.4 billion in payroll. The primary raw input to this sector comes from the Northern Forests. In various parts of the Northern Forest counties, the forest products industry constitutes between 20% and 65% of the manufacturing sector.

Agriculture

Although agriculture is not a major employer in the Northern Forest economy (less than 1% of employment), it does account for a large fraction of the land area in the Northern Forest counties. According to the 1987 *Census of Agriculture*, there are 14,774 farms that produce $460 million of sales in the Northern Forest counties. Agricultural land is 27% of the Vermont counties and 20% of the New York counties. Overall, 12% of the land area in the Northern Forest counties is in agriculture. Although the percentage of agricultural land in the Maine region is relatively small, these farms account for about one-half of the number of farms and agriculture sales in Maine. In total, the Northern Forest counties agricultural sector amounts to 28.2% of the farms, 13.8% of the sales, and 32.8% of the total agricultural sector across the four states.[9]

Recreation and Tourism

The Northern Forest's resource base provides enormous recreational opportunities. In Maine, New Hampshire, and Vermont, there are approximately 150 parks and recreation areas, 120 wildlife management areas, 115 trails for hiking, 105 sites for canoe trips, and almost 400 campgrounds.[10] The Adirondack Park in New York offers another 9,400 square miles with more than 2,800 ponds and lakes, 1,500 miles of rivers, and 30,000 miles of brooks and streams.[11]

The skiing industry is one of the most important sources of revenue in the tourism sector of the Northern Forest states. Downhill skiing attracts

10 million skier days per year. There are an estimated 218 facilities that employ about 14,000 full-time and 10,000 part-time workers. The nordic skiing industry is relatively smaller with 1.5 million skier days. About 200 nordic ski touring facilities provide jobs for roughly 1,600 full-time and 1,600 part-time employees.[12]

Hunting and fishing is another large component of the tourism sector. The 1985 National Survey of Fishing, Hunting, and Wildlife Associated Recreation estimates there are nearly 2 million freshwater anglers that participate in 31 million fishing days. The same survey found 1.2 million hunters that engage in over 20 million days of hunting. Over half of the hunting activity is devoted to big game (deer, moose, bear, and wild turkey), another 30% for small game (rabbit, quail, grouse, squirrels, and pheasant), and 7% for migratory birds (geese, ducks, coots, rails). These hunters spend about $160 million a year in the Northern Forest states.

The region attracts millions of other visitors, and additional income, for the fall foliage, hiking, camping, horseback riding, canoeing, river rafting, bird watching, snowmobiling, and other recreational pursuits. These recreation activities are a significant source of employment and income to the Northern Forest economy's tourism industry.

Tourism spending occurs through a number of sources in the economy. Vacationers typically stay in hotels, eat meals in restaurants, and visit special types of tourist or recreational outlets. Although it is not always possible to distinguish spending by residents relative to spending by tourists, tourism will be strongly correlated with these activities. In this study, the tourism sector is defined by combining the following sectors: hotels and other lodging, eating and dining places, and amusement and recreation services.

Tourism employment and payroll in the Northern Forest counties are summarized in Table 4.6. In total, the tourism sector provides over 50,000 jobs and a payroll of $455 million. Tourism accounts for 10% of total employment in the Northern Forest counties. For Vermont and New Hampshire, the share of tourism employment rises to about 18% of total employment. In contrast, the tourism payroll share is relatively smaller than the tourism employment share. For the entire region, the tourism payroll is 5% of the total payroll, which is half of the employment share, a pattern that is consistent across each state. This finding provides evidence that tourism-based jobs provide lower wages relative to other workers in the economy.

For 1987, it is estimated that out-of-state vacationers spent $1.2 billion in Maine, $1.1 billion in New Hampshire, $0.9 billion in Vermont, and over $16.0 billion in New York. These figures do not include expenditures

TABLE 4.6.
Employment and Payroll in the Tourism and Development Sectors in Northern Forest Counties in 1990

	Tourism sector		Development sector	
	Employment	Payroll (million $)	Employment	Payroll (million $)
Northern Forest counties				
Maine	13,050	105,772	9,773	224,014
New Hampshire	2,206	19,516	535	9,065
Vermont	5,539	43,358	2,309	46,628
New York[a]	31,806	287,165	17,517	443,614
Total	52,601	455,811	30,134	723,321
Percentage within Northern Forest counties				
Maine	9.3%	4.1%	7.0%	8.7%
New Hampshire	18.0%	9.2%	4.4%	4.3%
Vermont	17.5%	8.5%	7.3%	9.1%
New York[a]	10.0%	4.9%	5.5%	7.6%
Total	10.5%	5.0%	6.0%	8.0%

Source: Data from U.S. Bureau of the Census, 1990, *County Business Patterns*, Washington, D.C.: Government Printing Office.
[a]1989 data.

by in-state residents, and therefore understate the actual level of total expenditures for the region.[13] In summary, these figures show that tourism is the second most important source of employment in the Northern Forest economy.

Development: Real Estate and Construction

The recreational opportunities and aesthetic appeal of the Northern Forest region provide an attractive location to purchase a vacation home. To meet this growing demand, development forces in the region provide employment in real estate and construction. Table 4.6 demonstrates the magnitude and relative contribution of employment and payroll for this sector. In the Northern Forest counties, there are 26,200 jobs in construction and nearly 3,900 positions in real estate. The combined total of 30,100 jobs in the development sector amounts to 6% of the total employment in the Northern Forest counties. The development sector payroll of $723 million contributes 8% of the total payroll in the Northern Forest counties. Although the development sector provides 22,000 fewer jobs than the tourism sector, the development payroll is 158% of the tourism sector payroll. Thus, the development sector jobs are on average higher paying than tourism-based jobs.

Summary

This overview of the Northern Forest economy yields a number of key findings. The Northern Forest counties contain 1.8 million people and 500,000 jobs. About 50% of all employment is in manufacturing and services. Within manufacturing, the forest products industry provides 100,000 jobs statewide, primarily in the Northern Forest counties and in the paper and allied products industry. Within services, the tourism industry creates 50,000 jobs in the Northern Forest counties. Agriculture contributes less than 1% of the workforce, but utilizes 12% of the land. And finally, the development sector provides another 30,000 jobs to the Northern Forest economy.

Economic Transition

The forest products industry, tourism, and development are all important components of the current Northern Forest economy. The interesting issue concerns the direction of change in the Northern Forest economy. In essence, there is a question of land use between traditional forestry uses and development pressures. The 1988 sale of the former Diamond International forest land by Cie Generale Electricite raised public concern about the conversion of land for development in the region. The public reaction prompted the creation of the Northern Forests Lands Study and ultimately the Northern Forest Lands Council. The mission statement of the council is to "reinforce traditional patterns of land ownership and uses of large forest areas" by promoting sustainable forestry over large areas that simultaneously protects wildlife and other resources.[14] Environmental groups concerned about the ecological resources in the region would like to see more activist intervention to protect the forests. In contrast, some landowners and property rights activists are concerned about potential restrictions that may be imposed on land in the region. Given the different perspectives, land-use management policies are likely to be a controversial topic and the subject of considerable debate in the political arena. In order to gain more insight on this issue, this section takes a closer look at the economic forces driving development in the region and the long-run trends in forestry markets.

Development in the Northern Forest

Currently, the market is the front line in the battle over land use in the Northern Forests. Economic forces determine whether an acre of land is more valuable for forestry or development uses. Private landowners are motivated to put their land to its most valuable use. Considering that 84% of the Northern Forest region is held in private ownership, shifts in market forces can have a sizable effect on the land-use patterns in the region. During the late 1980s, land value for development was at a substantial premium relative to land used for forestry. As a result, the market encouraged private landowners to convert their land from traditional uses in forestry or agriculture to development. This means that land is typically subdivided into smaller lots and used to build vacation homes. Although the recent recession has dampened the real estate boom, one may suspect that an upturn in the business cycle would accelerate more land conversion in the region.

The Northern Forest counties currently contain a large number of vacation homes. The 1990 Census reports the number of housing units and vacant housing units for vacation or seasonal use. As shown in Table 4.7, there are 125,000 vacation housing units in the Northern Forest counties. For the Northern Forest counties of Maine, New Hampshire, and Vermont, vacation housing comprises almost 20% of the total housing stock. The inclusion of New York reduces the percentage of vacation housing down to about 15%. In contrast, the nearby states of Connecticut, Massachusetts, Rhode Island, and New Jersey have a significantly lower fraction of vacation housing, amounting to 3.1% of the total housing stock.

In the Northern Forest Lands Study, a case study of subdivision activity was undertaken in five Northern Forest counties: Franklin and Washington (Maine), Coos (New Hampshire), Orleans (Vermont), and Essex (New York). The study documented the amount of subdivided land in specific towns from 1980 to 1989. The data yield two interesting findings. First, the amount of land subdivided relative to the amount of land in the towns appears rather small. In three of the five counties, subdivisions accounted for less than 2% of the land in the towns studied. The two other counties experienced relatively higher rates of 6% and 10%. Second, the amount of subdivision land is rather large when measured against the amount of commercial forest land. Depending upon the region, subdivision land is between 40% and 60% of the nonindustrial private land. This suggests a much higher level of scarcity for the type of land most vulnerable to conversion.[15]

The demand for vacation homes in the Northern Forest is driven by

TABLE 4.7.
Summary of Statistics on Vacation Housing in the Northern Forest Region for 1990 with Comparative Data for the Southern New England States and New Jersey

	Total housing units	Vacant for seasonal, recreational use	Percent of total housing units
Northern Forest counties			
Maine	234,390	46,177	19.7%
New Hampshire	18,712	3,613	19.3%
Vermont	57,971	11,186	19.3%
New York	533,017	64,839	12.2%
Total	844,090	125,815	14.9%
Northern Forest states			
Maine	587,045	88,039	15.0%
New Hampshire	503,904	57,135	11.3%
Vermont	271,214	45,405	16.7%
New York	7,226,891	212,625	2.9%
Total	8,589,054	403,204	4.7%
Conn., Mass., R.I., N.J.	7,283,443	223,423	3.1%

SOURCE: Data are from the U.S. Bureau of the Census, 1990, *Census of Population and Housing*, Washington, D.C.: Government Printing Office.

individuals from outside the region, primarily from the large eastern metropolitan areas. These individuals are not just purchasing a second home, they are purchasing a bundle of environmental amenities that are linked to the home. The amenities include close access to recreation areas, open space, less crime, scenic landscapes, and other items. One of the most desirable characteristics for a vacation home is frontage property on a lake or river, which has commanded large premiums in the real estate market.

The demand for vacation housing or property is influenced by a number of other factors, such as income, price, and travel time. The New England economy, and that of much of the Northeast, experienced a boom period during the 1980s. For the states just outside the Northern Forest (Connecticut, Massachusetts, Rhode Island, and New Jersey), personal income per capita jumped nearly 32% from 1980 to 1990. This compares with a 19% increase in the United States as a whole.[16] Moreover, homeowners in Northeastern metropolitan areas derived large capital gains from the appreciation of their own houses during the 1980s. This enabled existing homeowners to receive home equity loans or second mortgages to finance new investments. By 1990, the median home value averaged over the same four neighboring states was $159,100. By comparison, median home value across the Northern Forest counties was $66,050. In other words, the median home value in the Northern Forest was less than 42% of the median home value in the nearby states.[17]

Travel time to the Northern Forest region is another critical variable in the demand for vacation homes. Shorter travel times reduce the cost of a household's trip to their vacation house. Currently, the Northern Forest is within less than a day's drive for many millions of people.[18] As new highways open and provide better access to the region, travel time is reduced, and the net benefits of owning a vacation home in the region increase.

Trends in Forestry

The future of forestry in the Northern Forest region will also depend on the state of the forest products market. If forestry becomes more profitable, landowners have less of an incentive to convert their land for development purposes. In this light, it becomes relevant to look at recent trends and long-run forecasts on the market condition for forestry.

The future of forestry in the Northern Forest region and the United States should be viewed in the proper context of the broader global market for forest products. The world contains just over 4 trillion hectares (ha) of forests and wooded land. With annual harvests of 3.4 billion cubic meters (cm), about one-half is used for fuel and the other half allocated for industrial products like lumber and paper. Although the United States holds only 7% of the world's forests and wooded land, it accounts for 25% of the industrial roundwood production. The other major producers of industrial roundwood are Canada, the Nordic countries of Europe, and the former Soviet Union.[19]

Within the borders of the United States, the South and the Pacific Northwest are the largest regional producers of forest products. The South and West account for 75% of the total forest land. In contrast, the four Northern Forest states of Maine, New Hampshire, Vermont, and New York contain only 6% of the U.S. total forest land. When measured against total timberland, the northern states contribution rises to 9%. Timberland is defined as forest land that is producing or capable of producing industrial wood and is not subject to governmental restrictions on timber use. Although relatively small in the domestic market, Northern Forest producers do have a regional advantage in terms of being located near the major metropolitan areas of the Northeast. The proximity to markets reduces transportation costs, and to some extent provides some insulation from other regional markets.[20]

Essentially, the Northern Forests are linked to the larger economic forces that influence the forest products industry. Since 1978, total sales have risen more than 100% to $90 trillion. The wood and lumber industry sector is rather cyclical because it is closely tied to construction activity for new housing, which is sensitive to interest rates and the overall state of the economy. The paper and allied products industry also feels the changes over the business cycle because firms will use more paper when business is booming and cut back on paper during the bad times. As a result, the industry experienced a slowdown of activity during the recession periods of 1981–1982 and again in 1991. From 1985 through 1989, the forest products industry enjoyed a high growth in sales and record profits: $6.2 trillion in net profits during the peak years in 1988–1989.[21]

During the expansion in the late 1980s, the industry followed the Wall Street trend of increasing leverage and consolidation through mergers and acquisitions. For the industry, the ratio of long-term debt to equity rose from under 50% in the late 1970s and early 1980s to more than 80% by 1991. In theory, the structural reorganizations encourage managers to streamline operations and make the firms more efficient. To critics, this is precisely the type of financial maneuvering that caused some companies to sell their forest land holdings in the Northern Forest region. At this point, there is not sufficient evidence to make a conclusive case that debt was an important factor in the land conversion process. There is some preliminary evidence showing a correlation between production levels of the major firms in the industry and the firm leverage ratio.[22] Even if a correlation exists, there are two points to consider regarding the relationship between debt and land conversion. First, if managers of these firms suddenly realized they were holding forest land that had a higher value for development uses, the firm would be maximizing profits by selling the land to developers independent of the firm's leverage ratio. A sale would be in the stockholder's interest, although arguably not in society's interest. Second, if such a sale were counter to the social interest, the cause of the mistake is not due to the firm's debt structure, but the inability of the market to price these resources consistent with all social benefits and costs.

In looking at the future of forestry, Roger Sedjo and Kenneth Lyon provide a comprehensive analysis of the world timber market.[23] They constructed a complex economic model of timber supply and derived forecasts about the timber market over the next 50 years. The model was designed to reflect the characteristics of seven distinct regions of forests, the transition from old growth to secondary growth forest, and the level of investment in regeneration. Sedjo and Lyon present a base case scenario reflecting their judgment of the most reasonable set of assumptions. The base case forecast predicts a very modest 10% increase in timber price over the 50-year time horizon, or an annual price increase of less than 0.2%. This conclusion is counter to what they describe as the conventional wisdom of most foresters. The conventional wisdom assumes that demand will outpace world supplies, causing significant increases in the real price of timber.

Sedjo and Lyon examine alternative scenarios by changing key assumptions in the model. Two key findings warrant elaboration. First, changing demand assumptions alter the future price path. If the base case demand growth assumption of 1% per year is increased to 2% per year, the annual increase in market price accelerates from 0.2% to 1.2% per year. On the other hand, if demand slows to 0.5% per year, the market price becomes stable over time. Second, the future price path is also rather sensitive to assumptions about the introduction of higher supplies of timber to the world market. In particular, Sedjo and Lyons find that the development of new forest plantations could have a significant impact on supply. Also, higher than expected levels of production from the former Soviet Union over the next 20 years could further dampen the long-run price path. These new supplies moderate the growth of timber prices and offset the level of investments into regeneration in other regions.

Overall, the Sedjo and Lyon analysis indicates a supply of commercial timber that is rather responsive to the market price. As a result, there will not be rising scarcity of commercial timber and the long-run growth rate of price will be rather moderate. This implies corresponding moderate changes in stumpage prices of Northern Forest trees. If this analysis is correct, we should not expect the future timber market to radically alter the terms of trade between land valued according to its forestry use and development use.

Environmental Economics and Land-Use Policy

As humans place more and more demands on the Northern Forest resource base, a fundamental question is, how do we reconcile society's conflicting wants for these scarce resources? This is the basic problem of economics: Scarcity forces society to make choices. Economics provides us with guidelines. In general, an efficient allocation of resources produces the greatest benefits in excess of costs. In other words, we seek an allocation that maximizes net benefits. Benefits are broadly defined to account for all social benefits that resources provide members of society. Costs are defined in terms of opportunity costs. That is, the value of a resource is based on its next best alternative use. Given the proper valuation of all resources, the allocation problem becomes a matter of utilizing resources that maximize net benefits for society.

The market is one mechanism to solve the resource allocation problem. The market is a decentralized decision-making process where individuals buy and sell commodities. The market permits individuals the freedom to pursue their own self-interest. In this perspective, the individual consumer is assumed to be the basis for determining social wants. Competition in the market forces entrepreneurs to make commodities at the lowest cost. Owners of resources, such as land or timber, advance their own self-interest by practicing good stewardship and offering these resources to the most valued uses. The market functions by allowing price to equilibrate the forces of supply and demand, a phenomenon described by Adam Smith as the workings of the "invisible hand." Price plays an important informational role as an indicator of scarcity. Thus under ideal conditions, the market functions as a mechanism that channels resources to the most productive and socially beneficial applications.

Markets can also fail to attain the social optimum of maximizing net benefits. A. C. Pigou first recognized that spill-over effects, or externalities, cause the private market allocation to differ from the social optimum.[24] Externalities can generate benefits (positive externalities) or damages (negative externalities) in society. Consider the case of a negative externality where an electrical utility emits sulfur into the atmosphere while producing electricity. Winds may transport these pollutants into the Northern Forest region and subsequently damage the lakes and trees. Acid deposition in lakes kills the fish population, reduces the recreational benefits to anyone visiting the region, and diminishes the property value of landowners. These are real social costs imposed on society. In this sense, the production of electricity generates a negative externality. The market price of electricity fails to capture the damages imposed on the users of these resources, and therefore understates the true social cost society bears in producing electricity. If the utility were forced to face the full social costs of production, it would have an incentive to find cleaner production technologies, reduce output, compensate damaged parties, or some combination of these.

Sometimes the externality problem can be resolved by a market solution. Ronald Coase showed that parties affected by an externality will find it in their own interest to negotiate a mutually beneficial arrangement.[25] These solutions effectively internalize the externality and ensure that resources are allocated to their best use. The Coasian solution will only occur, however, when there are low transaction costs in negotiating agreements. As the number of affected parties increases in an externality problem, transaction costs become quite high, and the incentive for individuals to reach a settlement is reduced. This is particularly relevant for environmental externalities that are typically widely dispersed to the general public. Efforts to mobilize individuals for a collective action are also hampered by the free rider problem. This means the individual has a strategic interest to minimize personal sacrifice in making contributions for a public good, while deriving the benefits from the collective action.

The preservation of areas containing unique natural resources poses an interesting economic problem. In a region with a forest ecosystem, society derives valuable commercial and noncommercial benefits from this resource. How should society choose between competing uses, such as forestry, development for vacation homes, recreation, or aesthetic preservation? At one extreme, Milton Friedman advocates letting the market decide. He argues that market forces will determine the efficient level of land set aside for preservation and recreation, not the government. "If the public wants this kind of activity enough to pay for it, private enterprises will have every incentive to provide such parks."[26] In contrast, others argue that inherent externality problems cause the market to fail in preserving unique environmental resources. Burton Weisbrod notes that some individuals may not express their demand to visit a region today, but would be willing to pay to preserve their option to visit an undeveloped region in the future.[27] This is known as an option value. Additionally, John Krutilla suggests that some individuals derive satisfaction from the knowledge that a unique natural resource exists, even if they have no intention of seeing this resource in person.[28] This is called an existence value. If the market omits items like the option value and existence value, then it will yield an inefficient and inadequate amount of preservation of natural resources.

Policymakers confronted with a question of land use and the environment should consider basic principles. If they find markets functioning well, then the appropriate policy is a laissez-faire one. On the other hand, if policymakers determine there are serious market failures, then government action becomes justified. One route is for government to own and manage the land in question. Alternatively, government may allow private ownership but impose restrictions on the types of uses in a region through zoning or other regulatory mechanisms. Zoning can be designed to minimize the problem of externalities from one person's land affecting the neighbor's land. Despite considerable research and empirical work on the economic effects of land-use regulations, there are many unanswered questions about the efficiency of zoning. William Fischel offers the following four maxims about the current state of knowledge on the economics of zoning.

1. Local established land-use regulations (zoning) must be viewed as a flexible and decentralized network of restrictions, not a single-valued constraint on all building activity.

2. Zoning confers both benefits and costs that are capitalized as increases or decreases in property values.

3. Zoning is the product of economically rational political activity.

4. We do not know much about the efficiency of zoning, but aggregate community land values may be the key to measuring it.[29]

The last maxim provides policymakers with a criterion to judge proposed land-use policy. Does the proposed policy preserve positive externalities (or restrict negative externalities) that outweigh the costs of restricting specific land uses? If the answer is yes, then the proposed policy improves the allocation of resources for society. If the answer is no, then the proposed policy reduces economic efficiency.

5

Ethical Tensions in the Northern Forest

Stephanie Kaza

Forest as home, forest as resource, forest as landscape, forest as community, forest as wilderness—there is no one way to view the Northern Forest. Each person, town, corporate landholder, tree, moose, and salamander has a different perspective of and experience with the Northern Forest. In the struggle for biological and economic survival, some needs are incompatible. Some organisms and landscapes flourish while others perish. The ethical dilemmas of the Northern Forest arise out of the urgent and motivating desire to survive. Loggers, forest towns, native peoples, paper companies, and the forest itself all have this desire to survive in common.

The conflicts of the Northern Forest are not unique to this biological region. Similar tensions wrack the great conifer belt of the Pacific Northwest, the abundant rainforest of Amazonia, and the oak woodlands of California.[1] They are driven by the overwhelming imbalance of power of humans and human institutions over nonhumans, that is, trees, deer, fish, watersheds. This crushing imbalance raises many moral and ethical questions—What is right and wrong in each particular situation? There is no single answer. Each conflict is far too complex to be reduced to simple moral prescription.

In this chapter, I outline five ethical polarities, pointing out the tensions in each. The pressing situation of the Northern Forest has raised concerns to the point of threat to survival on many fronts. Under the stress of this struggle, as in all environmental conflicts, people tend to take sides, thereby freezing dialogue, cutting off the possibility for genuine moral engagement with each other. These tensions will not likely disappear, no matter what ultimate solutions are agreed upon. They reflect a long history of relations between forests and people and all the political, economic, social, and psychological struggles for power and definition that have determined the fate of the forests over many not-so-peaceful centuries.[2] How these dilemmas are resolved for the Northern Forests will reveal the moral intentions of the current players. It is my hope that this chapter helps to clarify the tensions and serves to find a way forward that will make sense to the generations to come.

Simplification Versus Complexity

Environmental conflicts are almost always too complicated to manage easily. In an attempt to control the uncontrollable, the human mind often opts for simplification. This tendency can be seen in ecological, economic, and social realms, all of which affect communication and problem solving in the Northern Forest. In contrast, the natural tendency of systems and life forms is to evolve toward complexity. Tensions arise over what stabilizes a system and what destroys or enhances the forest.

Ecologically, the tendency to simplification can be seen in the homogenization of the landscape. Diverse New England forest habitats, once reflecting a checkered history of storms, glaciers, invasions, and migrations, have been altered extensively by successive waves of deforestation, agriculture, and urbanization.[3] Where succession has been disrupted or set back by major natural or unnatural disturbance, the forest stays in an early stage of colonization. Clearcutting reduces a forest to a field, encouraging growth of pioneer plants and aggressive invaders. Changes in microclimate impact soil health and stability, tree growth, and wildlife diversity. Although simplified systems sometimes show temporary increased species diversity, these are only successional species, not the full array of species in a stable ecosystem. In general, human activities of the twentieth century have tended to simplify ecosystems and reduce complexity.

The tendency toward complexity is articulated by those trying to save endangered species or preserve habitat variety across the latitudes of the Northern Forest. Ecological complexity is appreciated by those who understand the details of mycorrhizal root associations or composite acid deposition. Those who value complexity tend to be college-educated resource professionals or environmental activists. Their primary activities in this regard are stopping the acceleration of simplification and designing restoration programs to reintroduce complexity. The urgency of their vision can be offensive to those who explain the forest in simpler terms.

Economically, simplification is well established by current protocols omitting external costs of production. Logging on Northern Forest lands by large timber or paper companies generates additional local costs in utilities, social services, and bureaucracies for forest regulation. Many of these costs are borne by the public. Paper manufacturing and wood products processing generate air and water pollution that affect lakes, rivers, and human health. Breathing bad air and drinking bad water cause increased health and social welfare costs, almost none of which are borne by the originating factory. For decades, it has been convenient for manufacturers to leave the external costs to the community to pick up, thereby increasing their own profits at the expense of the public. A more complex budget would include external costs, shifting the burden of accountability to those responsible.

Large corporations have tended to focus on single goals; foremost is making a profit for company executives and their shareholders. In managing their land assets for short rotations, timber companies have justified heavy cutting and been able to sustain a high profit margin.[4] If company goals are defined primarily by the profit goal, there is little room for the complexity of multiple goals that would serve the region, community, and forest as well as the corporation. Where large companies dominate the local economy as in a number of Maine communities, single-company towns are precariously reliant on a single employer. In the Northern Forest lands debate, the tendency toward complexity is represented by voices for economic diversification and local self-reliance.

Socially, simplification appears in the form of stereotyping and generalization. In the midst of user conflicts, it is common for people to associate certain points of view with simplified labels—logger, activist, local, flatlander, rural resident, corporation, developer, Native American. Generally this leads to some version of polarization, in which generic conversations replace genuine dialogue between those who are actually involved in the local piece of land. Emotional charges ride high on these stereotypes, and can lead to name-calling and even violence. In the Adirondacks, environmental activists have received threatening phone calls and endured mysterious attacks on personal property.[5]

Regions as well as people can be reduced to stereotypes, oversimplifying the complexity of interstate relations. For example, if Maine is perceived as holding back negotiations because of pressure from its large corporate landholdings, or Vermont is seen as inconsequential because of its small piece of the forest, this blocks creative problem solving that depends far more on individuals working together than on state stereotypes.

The social tension of simplification causes a kind of dehumanization of individuals, each of whom cares about the situation from a particular lived experience. No one likes to feel reduced to a label. Unfortunately, it is simpler to make assumptions about others than to actually get to know them and make an effort to hear their concerns. To accept complexity is to recognize the breadth and depth of people's suffering. Few have the deep wellsprings of patience for this.

By definition, complexity includes what we don't know as well as what we do know. The tension between simplification and complexity often carries a perplexing combination of self-righteous knowing and humbling inadequacy. In some cases, the outcomes of simplification are incompatible with complexity. You cannot clearcut a large area and expect to retain ecosystem stability. Those who want to control the outcome of a Northern Forest environmental issue may often opt for simple solutions, precisely because they are more definable and enforceable. They may paint a negative picture of the "opposition" as if it were a uniform enemy. It is far more challenging to recognize the many facets of what is not known that generate true complexity.

Present Versus Future

The tension of present versus future is the tension of weighing short-term gains against long-term gains. Next to the vivid and incessant demands of the present, it is hard to imagine the future to come. Yet the character of the future is determined by choices made in the present. Short-term gains are favored by those in business now, looking to make a profit and achieve economic survival based on current assets. If these are gone in a hundred years, someone alive today cannot assume personal responsibility for these consequences. So who does bear the consequences? Those who have not yet been born.[6]

Economic valuation of the present is based on the market discount rate, which calculates the value of present goods against their worth later in time. Almost always, goods or resources (such as forests) are considered to have their greatest worth at present. Because they are valued at less in the future, the economic incentive is to maximize gain in the present. This is a strong driving force behind the pressure on current resources. A big corporation such as Georgia-Pacific or Boise Cascade planning to maximize profit through wide-scale harvest of Northern Forest trees will gain more financial rewards for accelerated activity now rather than later. There is no counterbalancing financial mechanism to favor the future.

Economic activity such as tree harvesting involves certain risks to workers, companies, and the public. "Acceptable risk" is usually defined in terms of the present, based on what is known about the risk-generating source at the present time. Risk analysis can be applied to pesticide poisoning, leaking radioactive waste, acid deposition, for example. But who defines "acceptable"? Victims of insecticide sprays or polluted water are not statistics; they are real people affected by real risks. Very little is known about how hazards to human and ecosystem health may act synergistically in the future. What appears acceptable now may bioaccumulate over decades or generations, causing greater problems later. It is almost impossible to adequately define "acceptable risk" because of this. Thus, definitions tend to reflect what is expedient and favorable for those defining the term.

It is possible, however, to enumerate some of the legacies being prepared right now for future generations and gain some idea of what they will be facing. For example, in New York and Vermont, children of this generation's children will need to cope with acidified waterways and soil and cumulatively weakened tree roots that may cause forest deaths in years to come. They will inherit pesticide-impacted soils from spraying that has reduced the vitality of naturally occurring microorganisms. They will find dioxin from the use of paper mill sludge as fertilizer now traveling through the food chain. They will see even-aged stands of planted trees instead of complex forests. They may never see some of the insectivorous tropical migrant warblers whose populations have declined from habitat loss. They will inherit an extensive road system, fragmenting wildlife corridors and woodlands across the state.

The tension between present and future arises around the questions: Whose future? Who will continue to exist in the future? Whose lives will be safeguarded or enhanced by Northern Forest decisions? Small private landholders? Large corporations? Transnational companies? Indigenous nations? Children of Euro-American settlers? Sometimes the tension is phrased as: Who will give up what now to make the future possible for whom?

In an age that celebrates speed, it is particularly difficult to stretch the human imagination into the future across the slowness of years.[7] Speed of travel, of communication, of stimulation, of mechanical harvesting limit the attention span to the immediate. Fast camera cuts and quick sound bites shape the modern sense of reality, robbing people of the capacity to look deeply into the past or future. In the demanding effort to just keep up with the enormous pace of change, people are easily overwhelmed. The present, though complex, is still easier to deal with than the not-yet-formed future. But as the Northern Forests come under increasing population pressure from urban centers not far away, the future is being determined already.

The question is: Knowing what we know about this region, can we choose a more effective, planned-for option that might be sustainable into the future?

Universalization Versus Indigenization

Universalization is the set of processes involved in imposing a standard ideology, power structure, language, or set of cultural norms on an area or group of people. The tendency of universalization is to homogenize, to reduce difference, to establish common values and traditions.[8] With regard to the land, it is the tendency to treat one territory like another, assuming similarity as an operational foundation. Indigenization, in contrast, is the set of processes that reinforce evolution of culture specific to a local region. The tendency of indigenization is toward heterogeneity, diversity, and a range of cultural traditions and expressions. Indigenous cultures reflect the specific limits and histories of living in particular places.

In the Northern Forest, the tendency toward universalization is represented by consolidation of landholdings by large timber and paper corporations who impose relatively uniform management/harvesting strategies on these lands. Another expression of this tendency is the pattern of increasing development, especially for second family homes. To the extent that "flatlanders" or outsiders alter the local community structure when they visit or resettle in Vermont, for example, they tend to bring with them a more uniform urbanized set of values. This can create tension around environmental issues because of the differences in class, education, and wealth between urban immigrants and local residents.

Universalization is experienced differently by those in power who are imposing an ideology or system and by those who are resisting it.[9] Native peoples or locals in New York or Maine who feel their traditions threatened by the industrial growth economy share a feeling of powerlessness of the small up against the large. They also resent their patterns of culture and community being disregarded as insignificant. Those in power justify their activities with claims of benefit to local communities, assuming a trickle-down morality, that what is right for the dominant player is right for all those affected by it. But this is not always the case.

Similar tension results from perceived universalization imposed by government agencies or environmental organizations. This is part of the resistance to "green lining," as if a boundary marked the area that would then be under the control of the dominant power.[10] Fear of increased environmental regulation and taxation for local towns and small landowners reflects a fear of domination by universalization.

The contrasting tendency, toward indigenization, places more value on local self-determination and respect for particular cultures and biological regions. Fifth-generation Adirondack residents undoubtedly know the history and weather patterns of their specific valleys or hillsides far better than newcomers. Native Abenakis still carry alive in their own traditions a lineage of respect and knowledge of Lake Champlain and the Green Mountains. Indigenization takes time. An individual or community must accumulate years or generations of information and experience about how the forest behaves in the place they call home. This kind of information, especially in the context of a shared social culture, is what enables a people to become native to a place. Some have argued that only local people are adequately equipped to know their own local forests. Some have argued that only by becoming native can one come close to being a good land steward.[11]

The speed and scale of universalization easily overcomes the slow pace of indigenization. The resulting power imbalance is sustained by money and institutional weight. The tension of domination and resistance to domination is a major force around the globe in North–South environmental conflicts, free trade agreements, and transnational corporation invasion. It is no surprise that this tension should also be felt and expressed in Northern Forest controversies.

A subtext of this tension is the classic insider/outsider struggle. "Insiders" are traditionally those who have lived in a situation for some time and have established personal relationships with others based on position, turf, and respect. These involve recognition of power and seniority and a common social community. Insiders negotiate social relations by a set of understandings of personal accountability to neighbors or colleagues.[12] Insiders hold status in rural communities and also corporate or government bureaucracies.

"Outsiders" do not recognize the unspoken order of relationships established by insiders and are often perceived as being inconsiderate, ignorant, and aggressive in their attempts to dominate social or environmental relations. This perception complicates Northern Forest dialogue, leading to defensiveness and suspicion as stable insider social relations are thrown off by encounters with strangers or outsiders. Ironically, government and corporate staff "insiders" feel threatened by environmental activists and local community members "outside" their realm of operations, while local "insiders" feel threatened by environmentalist, government, and corporate "outsiders." The insider experience is akin to the indigenous tendency; the outsider experience often represents the universalizing tendency.

Commodity Versus Community

Forest as commodity is ruled by market forces, by a capitalist economic framework and the rules that sustain that system. Forest products, unprocessed logs, and pulp are moved around according to demand and economic efficiency. In a commodity framework a tree is reduced to a thing. Tree as object is processed to serve the growth economy, on the premise that more growth is better—more houses, more paper, more wood products, more consumption. The gross national product, in fact, measures the rate of growth, acting as a barometer for the scale of buying and trading.

Commodities are traded on the open market where the goal is to make as great a profit as possible. The forest products industry, for example, exports raw materials to Canada, Europe, and Japan where markets pay more for wood than local markets. "Free trade" flourishes when a company can minimize costs of production by access to cheap raw materials, cheap labor, and few environmental restrictions. When forest liquidation becomes too expensive in one state (or country), a company will leave to do business in another state. As Wendell Berry puts it, "The global economy . . . operates on the superstition that deficiencies or needs or wishes of one place may be safely met by the ruination of another place."[13]

For trees to be traded, they must be seen as objects, void of any relationship with humans other than economic. From this point of view, trees are board feet, biomass, or stumpage—objects of specific value on the open market. This view of relationship, however, is somewhat limited and does not account for other roles of trees in human lives.

Commodity implies ownership, and with ownership comes some sense of obligation to what is owned. But what is owned? One may value the trees but not the land, or the tradable options and income but not the forest. The sense of obligation varies substantially from landowner to landowner, depending on size of holdings, cultural traditions, location, and the owner's capacity to pay attention to his or her land. Absentee ownership may be the most precarious in terms of developing moral relations with the forest.

In contrast to forest as commodity is forest as community. Speaking of his native prairies, Aldo Leopold defined community to include "soils, waters, plants, and animals, or collectively: the land." For Leopold, "an individual [tree, person, animal] is a member of a community of interdependent parts."[14] To respond to the forest as community is to see oneself as a member of the forest community. In this experience of relationship, the forest is part of the fabric of the self and the culture, based on mutual bonding and influence over time. A forest as interconnected web of relations is much harder to objectify than a forest as collection of things.

Forest as community carries the implication of forest as home, the land where one lives. Local residents of the Adirondacks whose homes go back many generations place a strong value on home and community as center point of activity. The forest is the context, the backdrop, a key participant in self-definition of individuals and communities. In taking responsibility for forest as home, local people develop shared traditions and morals that reflect relational values.

Tied to home is a sense of belonging, that one is familiar and accepted where one lives. In the forest as home, the local dweller gains a sense of self in place, self in relation, self in community. This is often marked by appreciation and gratitude for being part of what one loves—a sharp contrast to being defined by what one owns.

The tension between forest as community and forest as commodity tends to be expressed by locals versus nonresident landowners, whether corporate or second home. Private small-scale loggers may speak for a middle ground between the two polarities, seeking community-based decision making while still using the forest for commodity trading. Environmentalists may speak up strongly against abuse of commodity but not necessarily be grounded in a local community. In advocating a public form of ownership, they suggest it is possible to care for a forest without necessarily living there. This augments the tension between the two opposite views. In some cases, it is not possible to maintain the forest as part of the community and also clearcut it as commodity.

Private Versus Public Rights

Of the five ethical tensions I have listed here, this is the one that receives the most public attention. However, I have listed it last rather than first, to show the other tensions also riding in Northern Forest conflicts. I believe the private/public rhetoric often carries the weight of these other tensions, confusing the basic principles with other factors. Many volumes have been written on this classic debate over the last several centuries[15]; I will only briefly cover some key aspects of this last polarity.

Private rights imply individual freedom to use one's land as he or she sees fit. In popular perception, this is held as central to the American way of life, as a primary determiner of human activity and moral choices. This sense of individual freedom is based in an ethic of noninterference: that whatever one wants to do is all right so long as it does not harm others. This social contract is based more in the patterns of social relations than environmental relations between neighbors. In the late twentieth century, however, it is now quite apparent that there is very little one can do environmentally to one's land without it affecting adjoining land in some way—silting the creek, poisoning the groundwater with pesticides, creating air pollution from wood stoves, fragmenting wildlife habitat by logging. Ecosystems are not easily bounded by property lines.

Private rights are associated with self-determination, privacy, and autonomy. In Northern Forest issues, resistance to environmental regulation represents a fear of losing one's freedom. From this perspective, "green lining" is seen as drawing a conspicuous boundary around a set-aside area, thereby threatening such freedom. As needs for privacy increase because of population pressures from the urbanized areas, more and more lines are being drawn at local scales to keep people out of the forest.

Private rights also imply the right to be protected from unwanted infringements or intrusions. Irresponsible corporate management practices that create pollution or degrade the biological health of the land interfere with the rights of the local private landowner to environmental safety and stability. These rights, however, tend to be deemphasized in the current debate.

Landowners are not a uniform group across the Northern Forest. They fall into at least four types: (1) small, local landowners who live year-round on their land; (2) small landowners who use their land as a second, vacation home; (3) large corporate and private family landowners who actively harvest and work the land industrially; and (4) large corporate absentee landowners who are keeping the forest as real estate for future development or trading. The two key aspects to these groups are size of holdings and degree of presence. Large-scale landowners represent more money and therefore more power. Small-scale landowners perceive themselves to be relatively powerless next to these Goliaths. The aspect of presence or absence has a strong influence on the nature of decision making. Local people feel the most powerless when a large absentee landowner suddenly liquidates its assets, as in the big Diamond International sale in 1988. In a single monumental action, it seemed as if presence counted for nothing—for local landowners, environmentalists, and state governments. This seems to run counter to the heart of the private rights philosophy—that human beings should carry some weight as individuals, that their presence should count in the world.

Public rights, on the other hand, represent another American tradition of land-use planning and zoning. Public rights derive from common property practices of European tradition, also used by Native Americans before European arrival. National and state forests protect shared watersheds and wildlife, promoting some sense of commonly held trust and protection of the land. Government agencies are assigned the job of serving U.S. citizens by watching over the air, water, and stability of soil and land. Included in this sense of public is the implication that one has the right to security, health, and happiness—not degraded by another's exercise of freedom.

The tension between private and public reflects a tension of independence versus interdependence. Traditional American heroes have placed great emphasis on independent achievement, whether in conquering the Wild West or surviving the cold northern winters. Interdependence requires cooperation and relying on others, which is unpredictable at best and far from central to the American way of life. Yet interdependence is closer to an ecological model of relationships in a forest community.

Other tensions in this polarity are the resistance to limits on one's freedom and the debates around who will define what is private and what is public. The desire to control one's fate can be overrun by those who represent the public good. Thus, the conversation of private and public often reflects a power struggle between those with and those without power. It can easily derail conversation about what is sustainable for the forest. Power relations are rarely dealt with directly in environmental hearings, yet they almost always affect the dynamics of any forest argument where one party feels helpless. However, this does not necessarily mean the debate is about private versus public; it may be more a debate about who has power (in both a perceived and real sense) and who does not.

Power Relations and Questions of Ethics

In reviewing these five ethical tensions, one overriding influence is power. Who has the power to define the issues? To define the forest? Power can be used to manipulate outcomes, to control local economies, to trade resources for livelihoods, to degrade once flourishing forest ecosystems. The Northern Forest controversies represent a historical and ongoing shifting balance of power in New England. Perceptions and responses to power dynamics are central in addressing all environmental problems. The Northern Forest is no exception.

Power dynamics play out at the local community level, at the state and national levels, and within the global context. International environmental politics reflect the pressures of increasing population, a shrinking resource base, excess waste and consumption by the developed world, and a significant power and debt imbalance between the North and the South. State and national politics reflect increasing regionalization of environmental concerns against long-standing state traditions and characters. Local environmental politics suffer from widely diverse educational backgrounds and understanding of environmental issues. Often power players at one level overrule those at another.

Power may be defined in terms of size of landholdings and budget of organizations and corporate companies. However, power may also be expressed as personal presence—the capacity to stand behind one's principles, to speak from depth of experience, to reflect historical and traditional relationship to the land. The two basic types of power may be described as "power over" and "power from within." When these two come up against each other, there is not only environmental conflict, there is ethical conflict.

Trees, moose, trout, and blueberries do not have a voice in these ethical dilemmas. Rather, their lives are at the mercy of human beings whose industrial, economic, political, and social actions have great impact on the natural world. Nonhuman beings are clearly at the very bottom of the power ladder in the human configuration of things at present. Without language, rights, or capacity to communicate, nonhumans can make little contribution to solving environmental problems in the Northern Forest. Human beings are the only ones who can take moral responsibility for their actions.

It is my thesis that ethical deliberation is an extremely effective and appropriate tool for examining power relations. No matter what scale an organization, institution, agency, company, or community is operating on, one can ask—Is it accountable for its actions? Are the individuals in decision-making roles making conscious choices to act ethically with regard to plants, animals, and people? Raising ethical considerations can shed light on the five polarity tensions described earlier, offering a way forward when negotiations seem paralyzed.

I suggest here six guidelines for clarifying controversies within the Northern Forest. They provide one mechanism for challenging current policies and investigating alternatives. Taken together, they perhaps offer support for healing broken relations within communities and with the forest. While these do not address the specifics of environmental regulations or forest management, they do direct attention to those who must live with these decisions and their implications for others and for future generations. Each of these is a suggested direction to head, not a dogmatic prescription for problem-solving. Each includes the nonhuman, the ecological, the land community as well as the human. Each provides another way of framing the issues to include ethical dimensions profoundly significant to the way we choose to address the Northern Forest.

Toward Respect

Respect in the dialogue between users and dwellers in the Northern Forest across the various regions can only enhance communication and understanding. This might include examination of stereotypes and education as to the actual jobs and responsibilities of those who are engaged in Northern Forest issues. Do environmentalists have a sense of the lives of loggers?

Do absentee corporate landlords have a sense of the people who live in their communities? Do local residents recognize the native people who lived on the land before them? It seems important that forums for communication and problem solving err on the side of being inclusive, with effort made to hear usually overlooked points of view.

Respect for nonhumans requires some examination of the traditional hierarchies of value that place people at the top and soil at the bottom of the biological pyramid. By what ethical principles do trees have less standing?[16] What kinds of actions generate respect for forest systems, for forest inhabitants? What are the actual impacts of pesticides, recreational use, logging, real estate development on nonhuman beings? To engage in respectful relations with nonhumans one might err on the side of learning as much as possible about the details of their lives.

Toward Acknowledging Limits

Limits define a problem, an area in trouble, the finite capacity for a forest to respond to distress. The American tendency is to believe optimistically in the resilience of people and systems, as if there were no end to the amount of damage that could be done without serious consequences. The environmental history of the United States is a story of advancing and plundering across the continent, from New England to the Great Plains, across the Rocky Mountains to the California coast. The frontier mentality of the 1800s is no longer helpful in addressing the environmental difficulties before us. Naming the Northern Forest as a region is one way to begin to define the topics under consideration. Within this framework, it will help to be specific about the limits of each individual subregion—to understand the crucial soil differences between northern Maine and southern Vermont in supporting timber species, to trace the histories of Indian nations and white settlers in the Adirondacks, to see corporate real estate deals in terms of international financial limits.

In the human realm, acknowledging limits may help draw the line on abusive environmental practices that cripple local economies or impact community health. By recognizing shared limits, one may be more willing to work with others concerned about forest issues, despite differences of opinion and personality conflicts. One of the most frustrating aspects of environmental problem solving is acknowledging the limits to what is known about local ecosystems. In most cases, one wishes to know more in order to choose how to act. But often one must act anyway, in some degree of ignorance. One can choose to err on the side of humility and allow for what is not known, rather than making false assumptions.

Toward Restraint

As Aldo Leopold stated so succinctly, "An ethic, ecologically, is a limitation on freedom of action in the struggle for existence."[17] All ethics have at their root the principle of restraint. To choose not to kill someone, to not lie, to not steal—in each case, one is choosing to restrain from actions that would harm another. In Northern Forest controversies, people could choose to err on the side of restraint in developing timber cutting plans, in allowing second family home development, in name-calling at community meetings. To lean toward restraint generally allows more time for reconsideration, more options for the future, less damage to undo later.

Restraint has primarily been developed in the world's religious traditions as it pertains to the virtues of individual actions. But restraint in social actions has a place in the Northern Forest debate. Agencies and institutions can stand behind specific ethics of restraint that bear on their area of responsibility. Policies that restrict development from fragile lakeshores or creeksides err on the side of restraint in support of healthy ecosystems. Corporate labor practices that restrain accumulation of wealth at the top to promote economic well-being of laborers at the bottom extend ethical benefits to more people. Environmental groups who make an effort not to be self-righteous increase the possibility for other points of view to be heard. In each case, the group takes responsibility for its actions, cultivating a social ethic of restraint that allows for a wider conversation.

Toward Caring

Biologists often speak about the *carrying capacity* of the land or the forest ecosystem, referring to the number of trees or deer the area can support. A variation on this might be to examine human *caring capacity* relative to the land and communities that support human activities. The capacity to care is the capacity to take actions on behalf of others and oneself that sustain life and promote healthy development. To care for the forest is to take specific actions that sustain the life of the forest, whether that is staying on the trails, logging selectively, or putting up bluebird boxes. In the last 40 years, much human capacity to care has been derailed into consuming (shopping, buying, going to malls), an activity strongly encouraged by advertisers and product manufacturers. But this investment of time in material goods often carries less satisfaction than investing in human or nonhuman relationships.

To err on the side of caring is to choose to become a more caring person, one who is involved in the relationships that support one's life. In the Northern Forest region, this necessarily includes the forest and the forest animals and plants as well as the human neighbors. Caring takes time; care demands for family and community tend to take top priority. I am suggesting here that caring include time spent in environmental education, forest restoration, community organizing, and also time spent listening directly to the land, the trees, the birds, the mountains. To invest time in caring is to invest time in relationships—with neighbors as well as watersheds.

Toward Accountability

Being accountable is being responsible for the consequences of one's choices. This is fundamental to ethical or moral development. Economic accountability does not need to be an exception to this. Though it is arduous work, one can analyze in depth the economics of a corporation or a region to see whether it takes adequate responsibility for its actions. One major effort in this direction is Mitch Lansky's research for *Beyond the Beauty Strip*, a study of Maine's forest economy.[18] Is a corporation accountable for its labor practices? For its real estate transactions? For its environmental impacts? For the role of local industry relative to global transnational trading?

Ecological accountability implies erring on the side of more rather than less information and assessment of biological communities. Does a state government agency take responsibility for adequate biological monitoring of its forests? Is information available to citizens' groups and concerned environmentalists? Do the assigned agencies actually know what exists in the field that is represented on satellite data maps? Also, one can look at scientific methods used to assess current states of ecological habitats. Are they adequate as a basis for decisions? Are they influenced by who commissions them?

Professional accountability is another area to investigate for degree of care and attention in taking responsibility. Doctors and lawyers are expected to practice according to established codes of ethics, but most environmental professionals have yet to develop similar codes.[19] What does a forester consider good forestry? What does a hydrologist consider ethical hydrology? Those trained in the scientific method generally have little background in ethical issues and have not been taught to regard them as important to their work. However, with environments and communities under pressure, it may be necessary and appropriate for these resource professionals to show specific accountability for their work.

Toward Restoration

Much of the Northern Forest area has suffered from tourist development, unsightly clearcuts, acid deposition, and unplanned growth. Erring on the side of restoration can work to heal the impacts of harvesting and building on the land. Efforts to reverse acidification, erosion, and groundwater pollution, for example, are efforts toward system restoration, assisted by human restraint and regulation. To return the land and forest to a state of health where that health has been broken is good work.[20] This work brings joy to those who engage in it. There are ample opportunities for citizen groups to restore local creeks and watersheds, to clean up point-source pollution. In the effort to restore health, one invests care in a place, building a loving and satisfying relationship with the local land.

Restoring human relations that have been broken can also be helpful in generating cooperation. Euro-American settlers have work to do in this area with the original native inhabitants whose numbers are now so reduced. In regions torn by controversy over planning dilemmas, such as the Adirondacks, communities may need to restore relations around divisive issues. To err on the side of restoration is to stand for healing rather than harming. Harsh deeds must be acknowledged before healing can happen. But without some movement toward restoration, the damage carries a shadow on the land, felt by all who live there or pass through.

Conclusion

Each of these six directions offers possibility for forward movement from a foundation of ethical consideration. I believe that only ethical dialogue will fully address the power relations that control current Northern Forest issues. The web of interlocking interests, from industry to government, from state to federal agencies, from global economy to local landholdings, is complex and highly evolved. Power dynamics are not simply a matter of one group or individual pitted against another. Today's networks of interest groups working together can easily override cultural checks and balances designed to limit the power of any one group. Direct feedback loops of accountability are possible at the local level because of the scale of interactions and also because of the history of community values and responses to ethical transgressions.

But similar mechanisms of restraint do not apply at the global scale of activity. The links between interest groups are more important than the differences; being part of a network of power is often more important than being ethically accountable. International and national economic pressures affect Northern Forest real estate ownership and forest harvesting decisions. Since global networks of power are not geographically based in any given locality and are not necessarily committed to a local community, they are removed from traditional local restraints based on ethical accountability.[21]

Decisions affecting the future of the Northern Forest will be made by someone. The question is, by whom and how? To the extent that they are ruled by the most powerful voices, it is unlikely that the less powerful will have a full voice in the outcome. Lansky suggests a consensus model drawing on a coalition of representatives of all concerned interests.[22] President Clinton used a version of this model in his April 1993 Forest Summit in the Pacific Northwest.

I believe the six directions for ethical consideration will be best engaged with many voices speaking up. Social ethics regarding the environment are far from being well worked out. As natural landscapes come under increasing pressure for harvest, recreation, and development, environmental ethics are an increasing part of the dialogue. The ethical debate can only be expected to intensify. There may be no one solution that will serve all parties, especially if trees, birds, and wildflowers are taken into moral consideration. But the effort to examine the ethical implications of each decision will surely lead to greater care in living with each other and with the gifts of the land.

PART II

Views from the Public Sector

As is always the case when one examines the public policy process, the major actors come from both the public sector and the private sector. In this section, three actors from the public sector offer their views on the current state of the Northern Forest, the policy process, and what the future might hold in store. Carl Reidel, head of Environmental Programs at the University of Vermont, offers an insider's view of the policy process. Reidel was a member of the Governors' Task Force on Northern Forest Lands. As one of 12 members of this task force, Reidel is especially qualified to trace the evolution of the issue and the workings of the Governors' Task Force, the Northern Forest Lands Study, and the Northern Forest Lands Council. In addition, Reidel has many years of experience in conservation organizations—as a former president of the American Forestry Association and as a vice president for the National Wildlife Federation—and is trained as a forester.

John Collins is a fifth-generation Adirondack resident, has served on the Adirondack Park Agency board since 1984, and was appointed chair of the Agency in September 1992. In his chapter, he describes the Adirondack Park and how it works. This is of interest to the larger Northern Forest story for two reasons: first, because the park makes up nearly a quarter of the Northern Forest, and second, because the park is a working green line reserve, combining public land and private land regulation, that can offer instruction about the use of such reserves elsewhere in the Northern Forest.

The final chapter in this section presents the views from one local government in the Northeast Kingdom of Vermont. Local governments have an important role to play in this process, since it is these small towns that will feel the effects of changes in the Northern Forest most directly. Brendan Whittaker brings a wealth of experience to this issue. He is the chair of the Selectboard of Brunswick, Vermont, and has also been the Town Lister, Zoning Administrator, and a member of the Planning Commission. At the state level, he was Vermont's Secretary of Environmental Conservation and represents Vermont on the Northern Forest Lands Council. He is also heading the Vermont Natural Resource Council's Northern Forest Project. And lastly, he too is a professional forester.

These three views from the public sector—a view of the process, a view from a specific region, and a view from the local level—will help us to understand how concerns with the Northern Forest have been handled in the political system, to learn how one potential model for the entire region works, and to comprehend the concerns of those living in the Northern Forest.

6

The Political Process of the Northern Forest Lands Study

Carl Reidel

The story of the Northern Forest Lands Study began in the early 1980s and is a tale still unfolding in 1993. It is a story of a remarkable conjunction of dedicated and powerful people, brought together by unexpected events. It is a story of a vast and diverse forest region in which an equally diverse array of local people, corporations, governments, and organized special interest groups vie for power over the future. It is a story of a once-in-a-lifetime chance to redesign the way we do land conservation in America—a rare opportunity to reshape the way forest regions are managed and governed nationwide.[1]

Much of the story of the political process that led to the Northern Forest Lands Study and shaped its subsequent impact on events in the region remains unwritten. Many significant decisions were made or designed in private, informal meetings about which we must depend on the participants' memories to understand what took place. Those present do not always agree on what happened or even who played the decisive roles in determining what was agreed upon. Even the formal meetings of the Governors' Task Force were unrecorded beyond minutes of specific actions. Thus, the following account is my own perception of the events I witnessed and my interpretation of written records and recollections of others. I was a member of the Governors' Task Force, representing conservation organizations from Vermont, and a Trustee of the New England Natural Resources Center, and I attended most of the meetings of these organizations during the period recounted in this chapter. My recollections are incomplete at best, but hopefully accurate in essential elements. Perhaps the shortcomings of my view of events will provoke others to add their experiences and interpretations to the record of this fascinating and important story.

The First Warning Shot

The first warning that change was about to sweep the Northern Forest came in 1982 when Sir James Goldsmith acquired the Diamond International Corporation through a hostile takeover. Goldsmith was a "British-French entrepreneur [who] had established a pattern of seeking acquisitions that combined strong underlying assets with many smaller peripheral businesses that could be sold easily."[2] The acquisition included 976,000 acres of timberlands across northern New England and New York that Diamond International had consolidated from hundreds of individual ownerships over a period of 40 years.

Goldsmith sold most of the paper mill assets to James River Corporation, retaining the forest lands in a new company, Diamond Occidental Forest, Inc. Although most accounts of these events credit Goldsmith with putting these lands up for sale in 1987, he had actually sold both Diamond Occidental and Diamond International a few months earlier to Cie Generale Electricite (CGE), a European communications company. "Thus, in 1987, CGE was the driving force in selling these assets. Goldsmith . . . wasn't actually involved."[3] Nevertheless, it was he who fired the first shot in a chain of events leading to the launching of the Northern Forest Lands Study.

A Yankee Economist Fires Back

Although a few conservation groups were troubled by the Goldsmith takeover of Diamond International in 1982, most people in the region seemed unconcerned. The history of the Northern Forest was a continuing story of land consolidations and sales, including the recent acquisition of Brown Paper Company by the James River Paper Company and subsequent sale of some of its lands to Boise Cascade Corporation. Fear of massive subdivision sales seemed remote. The historical pattern for the industry was consolidation, such as the acquisition of many small tracts to create the near-million-acre tract acquired by Goldsmith in 1982.

Perry Hagenstein, a nationally known forest economist from Boston and Executive Director of the New England Natural Resources Center (NENRC), suspected something was changing in the fundamental economics of Northern Forest land ownership. Charles "Hank" Foster, former Dean of the Yale School of Forestry and founder-president of the NENRC, agreed, adding concern that Reagan's 1986 tax law would further destabilize past ownership patterns by eliminating favorable tax treatment of capital gains from timber sales.

With a small grant secured by Foster through the NENRC in 1986, Hagenstein took a careful look at the economic forces that were affecting the large forest land ownerships of the Northern Forest. In a concise report issued in the summer of 1987, he warned that: "Changes in ownership and use of the large forest holdings are already occurring, and more changes are likely. . . . The increasing spread between the value of this land for timber growing and for recreation and development puts pressure on current owners."[4]

Hagenstein's report explained what Wall Street and Goldsmith already knew, and what most foresters and many conservation leaders had failed to understand: "The large holdings, more than ever before, are viewed as profit centers by their owners and are expected to earn returns themselves that justify continued ownership." This, Hagenstein warned in appropriate Yankee prose, would "mark a sea change in forest ownership in northern New England. . . . It does not appear that forest land in northern New England has a positive value for timber production." Clearly, the value of these lands was increasingly as real estate for recreation, tourism, and subdivision.

While Hagenstein was preparing his report, the Trustees of the NENRC held several meetings with regional conservation leaders to consider what initiatives would be timely in response to his findings. Early in 1987, Paul Bofinger, President of the Society for the Protection of New Hampshire Forests, met with the NENRC Trustees to consider ways to protect public interests in the private lands of the Northern Forest. A seasoned veteran of land conservation in New England, Bofinger had long been advocating expansion of the White Mountain National Forest in New Hampshire and into Maine, possibly including parts of Vermont's Northeast Kingdom.

Bofinger did not, however, believe that the traditional approach of fee acquisition of large blocks into citadel National Forests would be appropriate in this region. Such acquisition would be painfully slow in the face of Hagenstein's "sea change," and it would likely trigger renewed debate over wilderness designations as experienced a few years earlier in the White and Green Mountain National Forests. Based on his experience in New Hampshire, he proposed a new model for National Forests in New England with acquisition of limited rights through easements and joint ownerships—a model that would soon be tested as events overtook the thoughtful deliberation of the NENRC.

A Second Shot

The ink was barely dry on Hagenstein's report when Goldsmith confirmed Hagenstein's predictions in late 1987. As described earlier, Goldsmith sold the lands to CGE, which then sought to sell them as quickly as possible in their entirety rather than by subdivision. Early bids by The Nature Conservancy to buy selected parcels were rejected, and lands in New York, Vermont, and New Hampshire were listed with a real estate firm, LandVest.

In New York, Lassiter Properties of Georgia bought 96,000 acres for $17 million through LandVest. The State of New York later purchased 40,000 acres of conservation easements and 15,000 acres in fee from Lassiter. The remainder is still for sale, with Lassiter now in Chapter 11 bankruptcy.[5]

In May 1988, Rancourt Associates of New Hampshire signed an agreement for the purchase of about 90,000 acres in New Hampshire and Vermont for $19 million. Earlier in 1988 the Society for the Protection of New Hampshire Forests and The Nature Conservancy attempted to negotiate a deal for these lands on behalf of the New Hampshire Retirement Fund for $16 million. But their letter of interest arrived while the Rancourt negotiations were in progress, and it was rejected by CGE.

In the face of this pending speculative sale to Rancourt, Bofinger moved quickly to organize the conservation community in order to negotiate a deal with Rancourt, who had made a $1 million nonrefundable deposit and probably needed cash to complete the purchase. Enlisting the interest and support of Governor John Sununu and Senator Warren Rudman, a series of hectic meetings were held with representatives of the state, U.S. Forest Service, Senate staff, and various conservation interests. Focusing on over 40,000 acres in the Nash Stream Watershed and some 5,000 acres of inholdings in the White Mountain National Forest, and under the pressure of time, a unique and innovative plan was devised for joint state and federal ownership of the Nash Stream Tract.

The traditional approach would have required an act of Congress to authorize a National Forest purchase unit by extending the White Mountain National Forest boundary, a lengthy process fraught with political pitfalls. Delaying action for such congressional action was out of the question. The Nash Stream plan called for immediate purchase of the tract with state and private funds, with later sale of easements to the Forest Service to finance part of the purchase and to create a new cooperative venture between the U.S. Forest Service and state of New Hampshire.

On October 27, 1988, 12 people representing the state, Rancourt Associates, the U.S. Forest Service, The Nature Conservancy, and the Society for the Protection of New Hampshire Forests met in Concord to exchange $12.75 million of state, federal, and private funds for a deed to 46,700 acres. A year later the U.S. Forest Service acquired a permanent conservation easement from the state of New Hampshire for the Nash Stream Tract. Thus was born a unique state–federal partnership in forest land management.[6]

This remarkable cooperative effort was political brinkmanship at its best, with an almost perfect cast of characters. With Paul Bofinger as the provocateur, Governor Sununu was the perfect match for Rancourt in the tough negotiations. Senator Rudman, working with Vermont's Senator Patrick Leahy, moved quickly to secure $5 million in federal funds. U.S. Forest Service Chief Dale Robertson supported the unorthodox easement arrangement and, with the support of conservation leaders, secured necessary agency and congressional support.

Unfortunately, the final cost of the acquisitions came close to what Rancourt had paid for the entire 90,000 acre tract. Had the state or federal government been in a position to form a Land Bank to acquire the entire tract earlier, resale of some portions might well have financed the cost of the conservation lands. It is a lesson to remember as we look to the future of public land acquisitions in the Northern Forest.

In Maine, 790,000 acres were sold in parcels to The Nature Conservancy, state of Maine, and Fraser Paper Company, or were retained by the James River Corporation and Diamond Occidental Forest, with most of the land remaining in forest industry ownerships.

By 1989, about 100,000 acres of the original Goldsmith lands in New England and New York were in some form of public ownership at a cost of about $25 million. Both Lassiter and Rancourt, who had purchased the lands for quick resale as residential and recreation tracts, were caught in a declining land market and are now in bankruptcy. Not everyone was pleased with some of the deals involving easements and limited rights purchases, but just about everyone knew that the Northern Forest would never be the same again. Hagenstein's "sea change" had begun.

Senator Leahy to the Rescue

While negotiations with Rancourt were intensifying in New Hampshire during the summer of 1988, a meeting in Vermont planted the seed of the Northern Forest Lands Study. Senator Patrick Leahy held his annual meeting with Vermont environmental leaders to discuss mutual interests and future legislative plans. At the end of a wide-ranging agenda I raised the Northern Forest issue, citing efforts by the NENRC and the Society for the Protection of New Hampshire Forests to warn the region of future problems like the Goldsmith takeover and the land sales to speculators Lassiter and Rancourt. Those present urged Leahy and his staff to seek emergency funds for a study and land acquisitions in New Hampshire. Leahy quickly joined forces with New Hampshire Senator Warren Rudman and Maine Senator George Mitchell, U.S. Senate majority leader, in a truly bipartisan political effort by New England partisans.

Mollie Beattie, at that time Vermont's Commissioner of Forests and Parks, quickly organized her counterparts in the other states and convinced Governor Madeline Kunin of Vermont to call on the governors of New York, New Hampshire, and Maine to join her in forming a special task force and to seek congressional support for a special study. Here again was an amazing bipartisan effort, especially given that Sununu and Kunin had exchanged some rough words across the Connecticut River in the recent past.

In one of the quickest and most cooperative actions by New England States since the Revolutionary War, a four-state Governors' Task Force (GTF) was created, with three members appointed by each governor. Within a few months Congress appropriated $250,000 for a special Northern Forest Lands Study to be conducted by the U.S. Forest Service in cooperation with the states. The newly appointed GTF was to serve as the "board of directors" to work with the study chief Steve Harper, former Green Mountain National Forest Supervisor.

The Governors' Task Force

Like the cast of characters working on the Nash Stream deal with Rancourt in New Hampshire, the 12 members of the GTF were a near-perfect team for the task at hand. Each governor had appointed a member representing state government, conservation interests, and large landowners. Beattie and her counterparts from Maine and New York, with the Director of New Hampshire's Land Conservation Investment Program, brought solid government experience to the table. New York's Ross Whaley, President of the SUNY College of Environmental Science and Forestry, and George Davis, Director of the Commission on the Adirondacks, brought a national perspective and considerable experience in land management and protection. The environmental conservation representatives were veterans of state and national organizations, and the three forest owner representatives were equally diverse in their backgrounds and current jobs. I was pleased to be a member of the GTF; this was a professional team of the first order, and a very astute group of conservation politicians.[7]

We functioned as a committee of the whole for most of our work, with ad hoc assignments to individuals as needed. The chair was shared by the government representatives, each taking the lead role as chair when meetings were held in the states they represented. Beattie, however, was clearly the GTF leader as she took on many of the special assignments and shaped the agenda of many of the key meetings. Her staff assistant in the Vermont Agency of Natural Resources, Charles Johnson, worked closely with Beattie and attended most GTF meetings. Perhaps most influential in the overall Task Force/Land Study effort, however, was the Northern Forest Lands Study (NFLS) coordinator, Steve Harper. His role as the only full-time, paid member of the GTF/NFLS "team" gave him a natural leadership advantage—a role he knew well from past experience.

The Northern Forest Lands Study

Steve Harper was the ideal person to head the Northern Forest Lands Study (NFLS). A veteran Forest Service professional, he had been Forest Supervisor of the Green Mountain National Forest during the creation of that forest's Management Plan—a plan often cited as the best produced on any National Forest in the nation. He had a proven record as an effective leader who could skillfully mediate between contending special interest groups on complex issues of resource management.

Harper was joined by two young Forest Service professionals, Laura Falk and Ted Rankin.[8] These three were the entire professional staff of the NFLS, and expanded their capacity by contracting with consultants for various studies and conferences. The enormity of their task is reflected in the sheer scale of the forest region they were to assess. Spread over 26 million acres, the Northern Forest is larger in area than all the national parks in the lower 48 states. It reaches from Maine's most easterly coast, across northern New Hampshire and Vermont and into the Adirondacks of New York. It includes almost 75% of Maine and a third of New York, by every measure a national heritage.

Not only were they faced with studying an immense area, but an immense area with amazingly diverse natural and human systems. That diversity created equally immense political problems. While the Adirondack Park represented one of the most daring attempts at public and private land-use regulation in the nation, Maine's North Woods was the citadel of fairly free-wheeling large private landowners, especially forest products industries in the unorganized townships. In between these vastly different social regions were the fiercely independent small landowners of Vermont's Northeast Kingdom and northern New Hampshire. Add to that the political differences among the four governors, especially between Vermont's liberal environmentalist, Madeline Kunin, and New Hampshire's irascible conservative, John Sununu, and you have all the makings for a lively debate. And lively it was!

It was this potential for political meltdown that forged an effective working relationship between members of the GTF and the staff of the NFLS. Initially, it appeared that the Forest Service's NFLS team would work independently of the GTF, which some felt would lessen the political risks. However, at the urging of Mollie Beattie and several GTF members, Senators Leahy and Rudman wrote the Chief of the Forest Service, Dale Robertson, essentially defining the role of the GTF as the "board of directors" for the NFLS. The issue was resolved and never again raised.

In retrospect, some of the more militant environmental critics of the NFLS process suggested that this wedding of the NFLS and GTF led to too many compromises and ineffective policy recommendations. They argued that the GTF should have played the role of devil's advocate, forcing the Forest Service to respond more directly to the concerns of environmentalists in the region. On the other hand, it is clear that the NFLS draft report would never have survived the public review stage if it had not had the unanimous backing of the GTF. The political savvy of the GTF was an essential element in the process leading to the final NFLS report.

From the outset, the charge to the NFLS from Congress was specific and was described in detail in the final NFLS report:

> In 1988, the U.S. Department of Agriculture, Forest Service was directed by Congress to study the timberland resources in northern New York, Vermont, New Hampshire and Maine and to *identify and assess*: (1) forest resources, including timber, fish and wildlife, lakes and rivers, and recreation; (2) historical landownership patterns and future ownership, management, and use; (3) the likely impacts of changes in the land and resource ownership, management, and use patterns; and (4) alternative strategies to protect the long-term integrity and traditional uses of land. Congress specifically stated that the study was not intended to lead to administrative action by any federal agency, but rather to provide Congress and the affected state governments with information for possible future action.
>
> Further, Congress directed the Forest Service to work with the Governors' Task Force on Northern Forest Lands to carry out this study . . .
>
> To clarify the direction for the Northern Forest Lands Study, Senators Patrick Leahy of Vermont and Warren Rudman of New Hampshire wrote the Chief of the Forest Service on October 4, 1988 stating, "The current land ownership and management patterns have served the people and forests of the region well. We are seeking reinforcement rather than replacement of the patterns of ownership and use that have characterized these lands."[9]

Understanding this charge is essential to understanding the political process leading up to the final report, and the final report itself. The key words are "identify," "assess," and "information." The key statement is the charge that the study should not "lead to administrative action by any federal agency." Further, there was no mandate to assess specific management practices on any lands, such as timber harvest practices. Some environmental groups were convinced that this was intended in part (2) of the charge and should be a central focus of the NFLS, believing it would reveal poor forest management on corporate lands and strengthen their case for public acquisition.

Even more important was the clear charge to seek "reinforcement rather than replacement" of the "traditional uses of the land." This infuriated some environmental groups even more than the nonaction clause, especially when the final report itself asserts that "Current trends are leading to a very different future."

In light of the specific charge from Congress, the NFLS focused efforts on a thorough description of the biological resources, patterns of ownership, inventory and mapping systems, and a broad range of strategies to reinforce traditional land ownership and use patterns. Heeding the warning "that the study was not intended to lead to administrative action by any federal agency," and to reinforce traditional uses, it did not consider seriously calls from some environmental organizations for the establishment of major new National Parks and National Forest expansions.[10]

Most of us on the GTF were also convinced initially of the need to find a new policy context for addressing the controversial, complex, and rapidly changing issues in the region. We believed that new context had to be an innovative public–private cooperative approach long advocated by Paul Bofinger and others. Neither the old citadel-style National Park model nor Adam Smith's arthritic invisible hand of the laissez-faire market were considered adequate to the challenge of the future. We were optimistic that some creative blend was essential and possible while also recognizing, as the NFLS Draft Report predicted, "the road will not be smooth."

The NFLS and GTF Tackle the Issues

The NFLS staff began work in October 1988, with monthly meetings of the staff and GTF in locations throughout the region. Others in the region were involved through two widely distributed newsletters and at public meetings held during the winter. Over the next 10 months dozens of experts and hundreds of citizens, foresters and wildlife specialists, land planners, and local and state leaders worked together on special reports,[11] at workshops and conferences, and at public hearings to assess the situation and consider the effects of alternative strategies. A major conference was held in May 1989 in Durham, New Hampshire, sponsored by the NFLS, Yale University, and the New England Natural Resources Center.[12] Harper and his associates, Laura Falk and Ted Rankin, traveled thousands of miles consulting with hundreds of people throughout the region.

The effort was fraught with controversy from the start, and not long after the GTF began regular meetings it was all too clear that we were grappling with issues that reached far beyond the immediate study. Some of us on the GTF did not believe that Congress' limitation on action policy recommendations in the NFLS applied to the GTF, and wanted public land acquisition and land-use planning high on the agenda as an integral part of any public–private cooperative strategy. Many environmental leaders in the region agreed and said so at public hearings, some of the more radical of them dressed as wild animals when testifying.

Equally adamant were representatives of private landowners. Most of the land under study was privately owned in a region that honors private property rights with Revolutionary fervor, a fact that Ted Johnson from the Maine Forest Products Council never let the GTF forget. He made it clear from the outset that any recommendation by the GTF of large-scale public acquisitions or regional green line–based land-use regulations, in which boundaries that delineated acceptable forms of land use were established, were unacceptable to Maine. "My principal task on this task force," he once told me, "is to be damn sure that neither a green line or any new federal reserves are recommended."

Other landowner representatives and some government representatives also preferred to limit discussions to economic assistance for private owners, limited state purchases of easements on critical lands, and other government incentives to maintain traditional patterns of private ownership and land use without major public acquisitions. But they also insisted that these incentives be linked to land conservation agreements with landowners.

Hoping to find productive compromise, the GTF and NFLS staff explored a wide range of innovative strategies: tax incentives coupled with landowner agreements on land use; landowner liability protection coupled with local land-use planning; the use of conservation easements and acquisition of development rights to protect critical environmental areas; and green line designations like those used in England and New York's Adirondack Park.

The wide-ranging exploration of ideas made for interesting discussions, but one hard reality could not be ignored. Private property rights had real value, and owners were unlikely to give up any of those rights for less than their real value. While recent sales had averaged about $250 per acre, the range was very wide, depending on tract size and many locational and site-specific characteristics. The complex problems of valuing rights exchanged for various incentives were formidable. Who decides and who pays? Are programs voluntary or mandatory? Can government guarantee continued tax incentives? While private land trusts and The Nature Conservancy had resolved these kinds of problems for specific tracts, the sheer scale and diversity of the region being studied compounded the problems enormously.

The single most controversial issue was probably the question of a designated green line—a specific boundary around the "Northern Forest" region within which special programs, incentives, and regulations would be focused. For advocates of large-scale public land acquisitions for national parks and forests, it seemed to be a reasonable alternative and an essential step toward gaining special federal appropriations and new incentive programs. For industry landowners in Maine it carried all the evil connotations of the Adirondack Park Agency boundary with its mandatory wilderness zones on private lands. The NFLS commissioned a special study of green line strategies [13] and held a special day-long meeting to familiarize the staff and GTF with the findings, but consideration of a Northern Forest green line never moved beyond academic discussions. Ted Johnson, the industry representative from Maine, was adamantly opposed to "any designated boundary, green, brown, or any other color." The major concerns were who would draw the line and who would make the rules and allocate federal dollars.

Meetings of the GTF became heated whenever discussions hit these bottom-line issues, with one meeting in Maine coming close to a meltdown. But GTF members backed off, falling back on old friendships and agreeing to limit the NFLS report to an exploration of alternatives. The GTF would consider policy issues independent of the NFLS after they assessed public reaction to the draft study report.

The Draft Report Sparks Response

The Draft Report of the NFLS reflected the GTF standoff. A wide range of alternative strategies was explained, more study was called for, and action recommendations were avoided. Following release in October 1988, public comment on the draft was sought over the next 3 months. Over 800 individuals and organizations expressed their opinions through letters and editorials, many coming from as far as the West Coast, and over 1,000 people attended 21 public hearings. It was clear that people cared deeply about the future of the region, and that they differed significantly about how that future should be shaped.

The public hearings reflected widespread frustration with the draft report. Representatives of one environmental action group, Preserve Appalachian Wilderness (PAW), labeled the draft "pro-industry" and "hostile to wilderness," calling for public acquisition of large areas for biological reserves.[14] At a well-attended hearing in Burlington, Vermont, professional foresters and representatives of the forest products industry denounced any considerations of public acquisition, forest regulation, or green line designations of private forests. Others spoke with passion as advocates of local control and private property rights, the first hint of strong opposition to the NFLS by these soon-to-be-called "wise-use" groups.

Most conservation groups cautiously lauded the report for providing, in the words of the Vermont Natural Resources Council, "a far-sighted common vision for the Northern Forest . . . founded on an appreciation and acknowledgment of the diverse, intertwined public and private interests involved." But few could agree on how to reach for that vision. That formidable task now belonged to the Governors' Task Force. The decision had been made by the GTF to work with the NFLS staff to revise the Draft Report in light of the public response, but to also issue a separate GTF Report that would make specific policy recommendations to the Governors and Congress.

The Final Reports Are Forged

At the first meeting of the GTF in 1990 everyone was braced for a fight. With the help of a professional facilitator from the Forest Service, the GTF began to forge a long "consensus list" of actions they would recommend as part of a comprehensive strategy. The list included most of the economic incentives and tax programs previously discussed, along with recommendations for "further study" of items that some members felt were too vague to endorse outright. But whenever there was an attempt to raise the topics of a green line or public land acquisition, other than for small tracts of significant natural areas by the states, positions hardened. Even though advocates of large-scale acquisitions expressed willingness to limit purchases to those where there was "a willing buyer and a willing seller" (thus rejecting any form of condemnations), some representatives from Maine would not discuss these issues further and they were never again seriously considered by the GTF.

On the issue of creating some sort of regional council to continue the work of the GTF, every attempt to move ahead failed. Maine representatives refused to accept a proposal for a Northern Forest Council put forth by the other states until "details were spelled out." Consensus on these key issues seemed impossible. Only a last-minute agreement to meet again in a few weeks saved the GTF from dissolution.

On March 1, 1990, the GTF met again to consider its final report and to make a last attempt at a compromise on the creation of an interstate organization. Every attempt to again consider a green line, recommend significant public land acquisition, or to establish an effective interstate commission was blocked by Ted Johnson, supported by Commissioner Ed Meadows of the Maine Department of Conservation. Under this clear threat from Maine to veto any such recommendations, the state delegations met independently to consider their next move. Vermont and New Hampshire discussed withdrawing from the meeting to form their own bistate effort, but reluctantly returned to the final session to seek compromise.

In the end the GTF caved in. Recognizing the necessity of keeping the four-state coalition alive in order to gain federal support, and the simple fact that most of the land area of the Northern Forest was in Maine, the GTF agreed to a proposal from the Maine delegation for a scaled-down version of the council. A "Northern Forest Lands Council" was recommended to the governors, with advisory and study authority to continue the work of the GTF for 4 years.

The report of the GTF mirrored the final report of the NFLS in its lack of policy direction. In the interest of keeping Maine at the table, the GTF failed to make the tough decisions essential to establishing the institutional and policy frameworks necessary to begin implementation of the best ideas in the NFLS report. There was no mention of direct federal acquisitions of public lands nor of green lines. The once-bright promise of the NFLS was lost in the gridlock of contending special interests. But the proposed Northern Forest Lands Council still seemed to be the last hope for some implementation of NFLS findings, even though it was defined in the GTF report as merely an advisory body "operated by the four states" with a 48-month life.[15]

Birth of the Northern Forest Lands Council

With the idea of a regional council barely intact in our final report, we again sought and got the help of Senator Leahy of Vermont to keep the Northern Forest effort alive. As chair of the Senate Agricultural Committee he was able to include authorization for continuation of the Northern Forest Lands Study in the 1990 Farm Bill and to gain appropriations of $1.075 million to support the proposed Northern Forest Lands Council (NFLC) and inventory work by the states and Forest Service.

Leahy also introduced the "Northern Forest Lands Act of 1991" to provide Congressional authorization and policy direction for the NFLC, including a role in land acquisition. Public hearings on the bill were as riotous as those on the NFLS draft report a year earlier, and a lot uglier. Property rights advocates had "discovered" the NFLS, labeling it another attempt to constrain property rights not unlike Adirondack Park Agency plans, Act 200 in Vermont, or the Endangered Species Act prohibitions on logging in the Pacific Northwest. They picketed hearings in New York, Maine, and Vermont. Their fierce opposition to Leahy's bill was summed up in a statement by a member of the Adirondack Solidarity Alliance: "We'll fight you in the courts, we'll fight you in the hills, we'll fight you in the valleys."

In the meantime, the soon-to-expire Governors' Task Force advised the governors to appoint members to the new NFLC and to move forward on the basis of congressional appropriations to the NFLC, a not uncommon way to claim authorization even though there was no legislative definition of the NFLC. Thus, the NFLC was born, albeit a bit illegitimate, and able to enter the debate over Leahy's draft bill.

The forest industry moved quickly to pressure the NFLC to offer an alternative to Leahy's bill. Ted Johnson once again took the offensive, arguing that Leahy's bill created "unwarranted federal intrusion" and that Leahy did not recognize the importance of maintaining "individual state control" over forest land management and conservation. The National Audubon Society, National Wildlife Federation, Sierra Club, and Wilderness Society responded, supporting Leahy's bill and calling for even more comprehensive legislation.[16]

Finally, in response to special interest pressures in Maine, and in recognition that there was no effective consensus between environmental groups and the forest products industry, Senator Mitchell of Maine withdrew his support of the bill. No further attempts were made to gain Congressional authorization for the NFLC.

The continuing pattern of political manipulation of the NFLS process by some forest industry leaders was all too obvious. Their early support of the NFLS and the GTF appeared to have been little more than a ploy to thwart park advocates, overzealous land planners, and potential takeover schemes for a few years. In that they succeeded. Later, they were effective in keeping any major policy initiatives by the GTF in the back court while the clock ticked, taking advantage of growing regional fears of a declining economy and rising emotional backlash from the "wise-use" movement. In the end they were able to permanently cripple any effective institutionalization of the NFLS report strategies or GTF policy recommendations by killing Leahy's proposed Northern Forest Lands Act. The conservation groups and moderate state officials were no match for the forest industry's political infighting skills.

The Northern Forest Lands Council Lives!

Discouraged but undaunted, the NFLC moved ahead, hiring a full-time executive director and staff, designating four state coordinators, and expanding council membership to four from each state (to include a "local" representative) and a representative from the U.S. Forest Service.[17] During 1992 the NFLC established several working subcommittees and issued bimonthly newsletters. It held nine "Public Input Sessions" from November 1992 to January 1993 "to gather public input on the issue areas being studied" by the subcommittees.[18]

The NFLC's published material makes it very clear that they understand their role, stating that the NFLC is "not a regulatory agency, . . . has no authority to regulate land uses or other activities, . . . is not a government agency, and . . . cannot and will not acquire land." It defines its task as an "advisory organization . . . continuing the work of the Governors' Task Force and the Northern Forest Lands Study through September, 1994."[19]

While the NFLC claims to be the heir of the GTF, its critics believe it has defined a much narrower role for itself, "to simply be a mechanism to reinforce traditional ownership patterns." So stated 19 member organizations of the Northern Forest Alliance (NFA) in a letter sent to the NFLC in response to the NFLC's draft 1992 Interim Status Report. The NFA urged them "to re-establish a comprehensive context for the Council's work for the coming year to address the full charge given by the Task Force, The Northern Forest Lands Study, and the Congressional Record."

NFLC Chairman Robert Bendick, a respected natural resources professional, is adamant that the NFLC is broad-based and will be able to set forth clear "policy recommendations and strategies to enhance the natural and economic integrity of the Northern Forest." He expects that the NFLC will be able to submit these policy recommendations to Congress and the states for action by the time the NFLC expires in 1994. The NFLC newsletter, *UPDATE*, sets the schedule as follows:

> The timetable for Council activities proposes subcommittee research end in May 1993 and draft recommendations development be complete by September 1993. The Council will distribute the draft recommendations by April 1994. The Council will then spend its final six months advancing its recommendations to the Governors and Congressional delegations . . .[20]

Time Is Running Out on the NFLC

Many conservation leaders doubt that the NFLC has the time or the will to tackle the complex political and economic issues that must be addressed to effectively implement any public/private cooperative model, especially in an emotional climate where adversarial positions are hardening. With its agenda heavily weighted toward data gathering, inventory, and mapping, some critics believe the NFLC is intentionally avoiding controversial policy decisions. Reacting to a talk by Mollie Beattie at a conference at the University of Vermont in November 1992 in which she urged the NFLC to "courageous and decisive action," NFLC Executive Director Charles Levesque said that he believed it would take "decades" to fully understand the problems facing the region and to reach some kind of common ground. For an organization due to sunset next year, Levesque seemed to confirm the conservation community's worst fears that time had run out for the NFLC.

With this sense of a regional gridlock over the NFLS and the NFLC, over two dozen conservation organizations banded together in a loose coalition: the Northern Forest Alliance.[21] With a list of members resembling a Who's Who of state, regional, and national environmental conservation organizations and a collective membership of over 10 million, the NFA could be a formidable force in shaping the future of the Northern Forest. But with an organizational membership including such diverse groups as the Society of American Foresters and American Forestry Association on the right, and the Environmental Air Force and Preserve Appalachian Wilderness on the left, common ground may be as elusive for the NFA as it was for the GTF!

The same issues that stymied the GTF haunt the NFA: the size and nature of public land acquisitions, the need and role of a green line, creation of some sort of interstate or regional institution with real authority, and the wisdom of such "hardball" tactics as using the Federal Energy Regulatory Commission (FERC) process for relicensing hydropower sites in Maine to extract land-use agreements from the forest products industry. If there is one area of agreement among the loosely affiliated members, however, it is that the Northern Forest debate must be elevated to the national level if there is to be any hope of breaking the regional gridlock and wresting power from the forest products industry.

In its first attempt to change the venue of the debate, 15 members of the NFA called for the acquisition of over 600,000 acres of the Northern Forest at an estimated cost of $100 million in a sort of letter to Santa Claus in December 1992. It is likely that additional members of the NFA and others will endorse this proposal as their governing boards have time for formal endorsement, opening a new chapter in the NFLS saga.

This could mark the beginning of a new drive for large-scale public land acquisition, and the beginning of the end for the public–private cooperative approach advocated in the NFLS and GTF reports. If so, it raises the question of whether the forest products industry was wise in blocking compromise efforts by the GTF, or if their tactics will prove to have been the fatal mistake that eventually leads to expansive new public nature reserves in Maine.

An Uncertain Future

I am increasingly persuaded that some form of aggressive public land acquisition may be the only viable option left open to regional leaders, for two reasons that are revealed from examining the history of the Northern Forest Lands Study and the Governors' Task Force—reasons why the present Northern Forest Lands Council, with its limited mandate, cannot achieve its intended mission.

First, there simply isn't time to resolve the complex property rights and land valuation problems necessary for the effective implementation of a public–private cooperative model. The primary mandate for managers of large corporate landholdings in the region is fiduciary responsibility to corporate owners. Altruism is not part of that mandate. As Hagenstein warned in his landmark report, these managers must treat these land assets "as profit centers . . . expected to earn returns . . . that justify continued ownership."

The vague economic incentives being discussed by the NFLC, dependent upon ever-changing government commitment in an uncertain economic future, will not deter corporate managers from liquidating these assets if necessary to balance the books. Far more research and analysis will be required to present these owners with real financial options than is possible with the NFLC's current resources. As I have stated elsewhere:

> The . . . agenda is awesome. Research is needed to untangle the American property rights bundle to help policy-makers value specific rights, so as to devise new ownership forms employing . . . easements, transferable development rights schemes, and creative trusts and leases. If innovative planning is to be acceptable to private land owners, large or small, we must find ways for them to capitalize on their land values without treating land as a single market commodity; just another form of pure investment, no more, no less. That will require a far more sophisticated understanding of the factors affecting land values, as well as ways to create markets for recreation, wildlife, heritage resources, and other nonmarket values of land ownership.[22]

There simply isn't time for such complex analyses, especially when any discussion of land-use planning and redefinition of property rights creates immediate hostile reaction from powerful special interests groups. Of even greater concern for me is the real possibility that many of the large forest products corporations are not as interested in staying in the land management business as they claim, and are looking to the lucrative profits open to them as land speculators themselves. This is not an unfounded concern, given their record of thwarting both the GTF and the NFLC in attempts to adopt any form of regional authority or comprehensive policy. If true, time has already run out for the defenders of the NFLS and the NFLC.

Second, it is increasingly clear that our federalist system of government in the United States will not accommodate the kinds of cooperative arrangements essential to implement regional strategies such as those suggested in the NFLS. The institutional barriers to intergovernmental cooperation within the states and between local jurisdictions, to say nothing of Constitutional requirements for interstate compacts, are formidable. Every individual party to a proposed intergovernmental agreement has absolute veto power unless there is a prior, binding agreement, such as in an interstate compact approved by participating states and the Congress. Given the present stance of powerful special interests in Maine, securing such an agreement is highly unlikely.

In summary, for these two reasons—time constraints and the limits of federalism—one might well conclude that cooperative public–private schemes are fatally flawed, and that major public land acquisition is an alternative worth serious reconsideration. Such acquisitions need not, however, be only in the form of block transfers of land into citadel national parks or forests. A combination of green lining and land banking is worth serious consideration. Following is a brief excerpt from a paper explaining this strategy:

> While many (public/private cooperative) techniques can be very creatively employed in a greenlining context, they are often fraught with complex valuation and legal problems. A possible route "around these mountains" is land banking. . . . This involves public acquisition of a substantial fraction of the land in a region for purposes of controlling future development. The land acquired is not committed to a designated use at the time of purchase or condemnation, but is reserved, leased, or resold later with various use restrictions in accordance with a regional plan.
>
> Governments in land banking are essentially in the real estate business, using the market rather than police-power regulation to shape ownership and land use patterns. It can be a creative way to employ such tools as easements, transferable development rights and other shared-ownership techniques, and a lucrative source of funds for the government . . .
>
> Public land banking could capture some of the profit and, thus, the financial resources that now drive the private speculative land development market. In a genuine green line reserve equity growth in land will belong, at least in part, to the (people of the) region.[23]

An added political advantage of a land bank–based acquisition program is that it meets most of the objections of property rights advocates to traditional land-use planning and single purpose public land acquisitions for designated reserves.[24]

Whether the future of the Northern Forest lies in regional cooperation or a nationally based public acquisition program, perhaps employing a creative land banking strategy, one thing is certain: enlightened, courageous political leadership will be essential. Whether that will come again from Senator Patrick Leahy of Vermont in his pivotal position in the Senate, from the Northern Forest Lands Council, or from a national coalition of environmental organizations like the Northern Forest Alliance is unclear. But if such leadership is not forthcoming soon, "we will be handing our future to the knights and knaves who know well the present market value of our heritage."[25]

7

The Adirondack Park: How a Green Line Approach Works

John Collins

In size, diversity, and ownership pattern, the Adirondack Park is unique in the United States. Its 6 million acres are one of the most significant components of the Northern Forest, one-fifth the total area of New York state. Within its boundaries are vast forests and some lovely farmlands, towns and villages, mountains and valleys, concentrations of wetlands, lakes, ponds, and free-flowing rivers, unmatched in the Northeast, sharing intermingled private lands and public forest. There are 135,000 permanent and approximately 70,000 seasonal residents in the Park, and millions of tourists pass through it annually. Taken together, its public, state-designated wilderness lands constitute the largest wilderness east of the Mississippi River and about 90% of all the formally designated wilderness in the Northeast.

The park celebrated its Centennial in 1992. New York's legislature in 1892 voted to designate an area within which to concentrate acquisition of land for the Adirondack Forest Preserve. Although the intermingled private lands were not legally included in the park until 1912, the area was christened the "Adirondack Park" and the boundary, expanded many times since, is called the "Blue Line."

The Adirondack Forest Preserve, created by the Legislature in 1885 and given protection in the state Constitution in 1895, has enjoyed more than a century of care and protection by the state. But the title of "park" for the far larger mix of private lands in the Adirondacks was little more than

This chapter draws heavily on material prepared by the staff of the Adirondack Park Agency, and to them I am grateful.

a convenient label until 1973. In May of that year, an act of the legislature transformed the region into a distinct and cohesive unit. The state's lawmakers passed the Adirondack Park Agency Act, the legislative embodiment of the Adirondack Park Land Use and Development Plan and Map, a plan that would ensure the sound management of the Forest Preserve and the wise use and development of private lands within the "Blue Line."

Thus, this element of the Northern Forest became the largest area in the nation subject to a comprehensive land-use planning and regulating program. And the Adirondack Park, for so long only a name and a blue line on paper, became a reality.

The custodian of this vision of the Adirondack Park, the Adirondack Park Agency, is an independent, nonpartisan agency within the state's Executive Department. It consists of eight private citizens, all New York state residents, appointed by the Governor for terms of up to 4 years, and three state officials: the commissioners of Environmental Conservation and Economic Development, and the Secretary of State. Five of the private citizens must be permanent residents of the park. Three must reside permanently outside the park, and not more than five may be affiliated with the same political party. The agency members are aided by a staff of specialists in planning, law, engineering, ecology, and economics.

The Adirondack Park Plans

Since one of the options being discussed for the Northern Forest is the green line approach, it is useful to examine how this approach works within the Adirondack Park "Blue Line." Of the 6 million acres in the Park, 2.6 million acres (43%) are state-owned, part of the constitutionally protected Adirondack Forest Preserve belonging to all the people of New York state. Private lands total 3.4 million acres (57%), devoted principally to forestry, agriculture, and open-space recreation.

In 1969, the Temporary Study Commission on the Future of the Adirondacks was appointed to consider the future of the Adirondacks. Acting on the Commission's recommendations, the New York Legislature established the Adirondack Park Agency (APA) in 1971 and instructed the APA:

1. To prepare, in cooperation with the Department of Environmental Conservation, a plan for the management of the Adirondack state lands to be submitted to the Governor; and

2. To prepare a plan that would regulate development on the non-state-owned lands of the Park.

Let us now examine these two plans.

Almost all state-owned lands in the park are Forest Preserve, protected by the "forever wild" clause of the state Constitution written in 1894. The Adirondack Park State Land Master Plan sets policy for the management of the state-owned lands in the Adirondack Park. As authorized by the APA Act, the plan is prepared by the APA in consultation with the Department of Environmental Conservation and approved by the Governor. It emphasizes both the preservation of wilderness values and the provision of wild forest recreation. First adopted in 1972, the plan is evaluated and revised where necessary about every 5 years. Sixteen tracts of Adirondack Forest Preserve, comprising slightly over 1 million acres, are designated wilderness. These areas are reserved for such wilderness uses as camping, hiking, canoeing, fishing, hunting, trapping, snowshoeing, and ski touring. Motorized equipment use and buildings are prohibited in wilderness areas.

Other categories of state land are primitive and canoe areas, managed similarly to wilderness areas; intensive use areas, where such uses as ski centers, public campgrounds, developed beaches, and boat launching sites are appropriate; and wild forest areas, the largest single category (totaling 1.2 million acres), where a variety of outdoor recreational uses is allowed, including the use of motorized vehicles in designated places.

Working within the framework of the State Land Master Plan, the Department of Environmental Conservation, in consultation with the APA, is charged with the preparation of unit management plans for each of the discrete units of state-owned lands. These unit management plans are based on more detailed levels of information regarding specific natural features and resources, and are intended to translate the broad policy directives of the State Land Master Plan into specific implementation programs for the day-to-day use and conservation of the Adirondack Forest Preserve lands.

The Adirondack Park Land Use and Development Plan applies to the 3.4 million acres of private lands in the park. Initially approved by the legislature, it is administered by the APA and provides for the participation of local governments. The plan is designed to preserve the natural resources and open-space character of the park while providing ample opportunity for appropriate development. The plan does this by directing and clustering development so as to minimize its impact on the park's natural resources, beauty, and open spaces.

The plan gives the APA regulatory authority over regional land uses, that is, uses of more than strictly local significance, and establishes standards that apply to all shoreline development in the park. Under the plan, all private lands are mapped into six land-use classifications. The classifications are based on many factors: existing uses and growth patterns; physical limitations relating to soils, slopes, elevations; identification of unique features such as gorges and waterfalls; biological considerations such as wildlife habitat, rare or endangered flora and fauna, and fragile ecosystems (swamps, bogs, and marshes); and public considerations such as historic sites, proximity to critical state lands, and the need to preserve the open-space character of the park.

For each classification, "overall intensity guidelines" prescribe the approximate number of principal buildings that are allowed in a square mile of the particular land use area, and a "character description" and statement of "purposes, policies and objectives" describe it. "Development considerations" point to possible adverse impacts of projects. Lists of "compatible uses" for each land-use area serve to guide development.

In addition, all projects requiring review and approval by the agency are listed. These "regional projects" vary with each area. In the case of subdivisions, for example, the agency's jurisdiction generally ranges from a 100-lot project in hamlets to a 2-lot subdivision in resource management areas.

Since the regulation of private property is of increasing political importance, both nationally and in the Northern Forest, it is worth examining these land-use classifications more specifically.

Hamlets. These are the growth and service centers of the park. Hamlet boundaries often include unsewered but densely settled areas like Warrensburg and Willsboro, and often extend well beyond serviced or developed areas to provide room for expansion. The plan permits all uses within hamlet areas and sets no limit on development intensity. (In all classifications, local governments may impose stricter land-use controls than those of the Adirondack Park Land Use and Development Plan.)

Moderate intensity use areas. Most uses are compatible, but relatively concentrated residential development is most appropriate. The overall intensity guideline in these areas is 500 principal buildings per square mile (1.3 acre average lot size).

Low intensity use areas. Most uses are compatible, but residential development at a lower intensity than just specified is most appropriate. The overall intensity guideline in these areas is 200 principal buildings per square mile (3.2 acre average lot size).

Rural use areas. Most uses are compatible, but rural uses are most appropriate. The overall intensity guideline in these areas is 75 principal buildings per square mile (8.5 acre average lot size).

Resource management areas. These cover nearly 2 million acres, or more than 50% of the park's private lands. Special care is given to protecting the natural open-space character of these lands. Compatible uses include agriculture and forestry, game preserves, and recreation. Residential development is compatible only at a very low density. The overall intensity guideline in these areas is 15 principal buildings per square mile (42.7 acre average lot size).

Industrial use areas. These areas are where existing industrial uses (including mineral extractions) are located and where future industrial development is most suitable. Additional areas may be identified by local and State officials.

Regional in scope, the Land Use and Development Plan provides a general framework within which the park's 105 units of local government can engage in local planning and zoning. Much of the development that occurs in the Adirondacks is local in scale and not subject to APA jurisdiction; it is important that local governments carry out their responsibilities to guide this development. The program has significantly evolved from the individual community approach to multicommunity projects, with implementation through circuit rider and other activities.

This land management partnership between local and state government was, until recently, supported by New York State through the APA's Local Planning Assistance Program. Technical and financial assistance was available from the Agency for consulting services to local governments by private planning firms, county or regional planning staffs, or the APA staff itself.

In preparing a local land-use program, community natural resources and social and economic conditions and trends are analyzed. Then a long-range plan and accompanying regulatory tools (zoning, subdivision regulations, a sanitary code) are prepared by the local planning board and thereafter officially adopted. With such a long-range plan, a community can indicate opportunities for new development while preserving its natural assets.

When such plans are adopted locally and meet the criteria for approval by the APA, jurisdiction over most regional projects becomes the exclusive responsibility of the local government. The APA also utilizes standards in the local plans in its review of those regional projects that remain its responsibility.

Seventy-eight towns and 13 villages have been directly assisted by the APA's Local Planning Assistance Program. The majority have adopted new local land-use regulations. Fifteen have received APA approval for these regulations, allowing local administration of most regional permit applications.

The APA Act provides that the Park Plan Map can be amended, provided that refined information and data indicate the proposed change will be in accordance with the purposes, policies, and character descriptions of the new classification, as stated in the act. These "map amendments" are based on whether the parcel in question fits the statutory character description for the classification sought, considering the natural resource characteristics of the area (soil, slope, wetlands, and other conditions), open space value, public access, and available public services. Potential patterns of use and other impacts that would result from the amendments are evaluated in an environmental impact statement prior to all but technical corrections.

Property owners may apply to the APA for amendments that result in either higher or lower allowable intensities of development. Local governments may also apply at any time for amendments to increase or decrease the intensity of development.

In addition to these land-use plans, the state legislature has designated over 1,200 miles of Adirondack rivers as part of the state's Wild, Scenic, and Recreational Rivers System. This designation preserves outstanding rivers in their natural, free-flowing condition so that they and their immediate environs will be preserved for the benefit and enjoyment of present and future generations.

Special regulations govern new development near designated rivers, and are designed to protect these waters from siltation due to erosion and pollution from sewage. They also prevent unwise and hazardous development of flood plains. The APA administers the Rivers System on private land within the park, and the Department of Environmental Conservation does so for rivers that flow through state lands.

Furthermore, the New York State Freshwater Wetlands Act is administered by the APA within the Adirondack Park. More than 14% of the Adirondack landscape is wetland, which serves a valuable environmental function by retaining flood water, acting to purify surface runoff water, and being generally excellent habitat for wildlife. Many unique forms of vegetation are also associated with wetlands.

Visions of the Future

In April 1990, 245 new ideas for the management of the Adirondack Park were presented by the Commission on the Adirondacks in the Twenty-First Century. A tepid to hostile reaction followed from Adirondack residents and others who found one or more recommendations not to their liking, or who criticized the lack of a structure for public participation in the translation of Commission recommendations into public policy and law.

The governor has since established a dialogue with representatives of local government and the state agencies principally involved in the future of the park. Various legislative proposals have advanced in different fora to address the major issues in the park: to reform the private land regulatory process, to overhaul the real property tax rules for private forest lands, and to revive the economic development process in the park. Some go further to effect fundamental reorganization of the institutions for park management, reassigning responsibilities variously to the Department of Environmental Conservation and the APA. None have advanced in the legislature due to opposition to any expansion of APA regulatory authority in the State Senate.

In this setting, the APA now struggles with three priorities:

Protect the economic integrity of the working forest. A complex question of fiscal (real property and income tax), acquisition (principally easements), and regulatory policies for which responsibilities are divided among many agencies in the state government (and federal taxing authorities). The APA oversight of subdivision of private forest lands is only a small part of this picture. Under the present system the APA's regulatory tools are only called into play when the larger system of public policies has failed to sustain the working forest.

Protect the ecologic integrity of the park's waters and wild lands. An issue that demands understanding of forces as diverse as air pollution causing acid rain and unregulated septic systems and land management practices that contribute phosphorus to degrade water quality in Adirondack lakes and streams. The APA's concern regarding the cumulative impacts of residential subdivision over which it exercises partial regulatory control (local governments govern subdivision in more developed Hamlet, Moderate Intensity, and Low Intensity Use areas) is central to management of private lands and shorelands. Similarly, the state land planning and related public use and access policies and responsibilities that are shared with the Department of Environmental Conservation determine whether critical elements of the Forest Preserve, such as the few acres of vegetation on alpine summits, will survive for future generations.

Protect the park's communities. This includes the economic and aesthetic values that make the region a marvelous place to live and raise families, as well as to recreate and visit. With only modest regulatory control in the developed Hamlets of the park and without financial assistance to local government planning, the APA has been hamstrung in this regard as well.

For the APA these pose many dilemmas. For instance, the Adirondack Park Agency Act seems to require a cumbersome technical analysis before advice regarding a landowner's need for a state permit can be provided. Often requiring many weeks, this basic procedural advice consumes technical staff time that might be devoted to guiding development rather than legal process. At Governor Mario Cuomo's invitation, I have recently appointed a small Task Force to search for nonlegislative solutions, and if they are not possible, to suggest changes to the law that would simply expedite the process, not change its substantive objectives. Coupled with modest additional funding provided for fiscal 1993–1994, there is hope that the cumbersome jurisdiction and permit determination process will become more responsive to the needs of its clients.

At the same time, the priorities just listed will require new combinations of agencies, authority, and financial resources to protect specific resources at risk. These combinations of institutions and tools should craft specific plans for forest lands, recreational greenways, and community development. With general place-specific guidance, a new effort to build consensus in and outside the park should then provide the financial and legal tools to make these plans real.

Resources are at hand. Creative combinations of state, local, and federal fiscal resources should begin an Adirondack greenways program. The federal Intermodal Surface Transportation and Efficiency Act (ISTEA) funds should be devoted in part to protecting and enhancing the scenic values of Adirondack Park roadways.

With a sense of urgency heightened by recent reports of financial woes of industrial forest landholders in the park, an easement program for forest land protection should also emerge from the thicket of controversy in the Adirondacks, based on elements that have been considerably refined by the Department of Environmental Conservation.

Here again are legal principles that divide the effort to bring agencies together and work for consensus to protect the Northern Forest. The APA is constrained by its regulatory role from engaging in advice on specific land acquisition proposals, including easements. However, the underlying economic, environmental, and fiscal information that should guide park policy has been assembled by the APA in the past, and this technical support needed by regulatory, planning, and acquisition programs continues to receive priority in the APA's planning program. These data sets require ongoing support and update.

What is also lacking is the institutional foundation and linkages to ensure public and interagency access to these policy and information resources over the long term. The APA looks to the Northern Forest Lands Council and the Northern Forest Resource Inventory project to continue to provide the forum and mechanisms for cooperation and public access that are so critical to success in meeting the urgent needs of the region.

The Adirondack Park as a Model for the Northern Forest?

This chapter has focused on how the Adirondack Park, a century-old green line area, works. Many of the issues that were raised here that are now occupying the energies of the APA and others concerned with the park are issues that affect the entire Northern Forest. My three priorities—protecting the ecologic integrity of the Park, protecting the integrity of the working forest, and protecting the Park's communities—are priorities throughout the Northern Forest. Among the more specific policies that have been debated are those dealing with tax reform, regulating private property, and stimulating economic development. Again, these are major considerations throughout the Northern Forest.

So, given these shared priorities and options, what can we learn from the experiences of the Adirondack Park and the APA that will illuminate the discussion on the Northern Forest?

Does the park work? Yes, and sometimes it works better than others. But today the Adirondack Park remains a vast area of private working forest, public preserve lands, and communities that has fared reasonably well economically in a period of overall decline in the Northeast (but with pockets of chronic and unacceptable poverty). It continues to be the bellwether for the long-term consequences of air pollution and acid rain, for which impacts can be measured for both disturbed and undisturbed ecosystems due to the foresight of stewards from previous generations. Extirpated species continue to return, some with help, some without. These are but a few of the many very positive indicators for the park.

As a green line park, the Adirondack Park benefits from a unique set of relationships between public and private ownership, responsibilities, and stewardship. I would emphasize the success of three core principles. First is the century-old commitment to "forever wild" public lands. The Adirondack Forest Preserve is a success story that provides context for the Park and its working forest. Second, for nearly a quarter century the Adirondack Park Agency has protected private open space lands, primarily the working forest serving the local mills and secondary wood products industry, and incidentally providing significant open space recreation uses. The subdivision of this open space for other uses is strictly regulated. Third, and sometimes most elusive from a regional perspective, vigorous communities depend on tourism, government, and nonforest and tourism-based entrepreneurs to provide jobs. Here state regulation is much less intrusive and most APA regulations can be replaced by locally administered rules.

These principles have stood the test of time, and most agree that they will continue. They, however, like much of the government, are far too complex. "Reform" efforts to date may have added to the complexity. Change that simplifies and elevates the core principles for better public understanding and more effective implementation is my long-term goal.

Most fundamentally, the Adirondack Park is a significant element of the Northern Forest, and its success with these issues is essential to the economic and environmental health of the whole region. As such it is its own model, only one of several in whose hands lies the future of the Northern Forest. Other notable examples include the Tug Hill Commission, the park's western neighbor, Vermont and New Hampshire's state, regional, and local authorities, and Maine's Land Use Regulatory Commission, offering state and regional counterpoints. We will all benefit from the dialogue over the region's future.

8

A View from Local Government

Brendan J. Whittaker

At 3:41 on Tuesday afternoon, March 3, 1992, Jim Bates, Town Moderator, banging down his gavel on an old teacher's desk in the small, one-room Town House, declared the Annual Meeting of Brunswick, Vermont, in session. The Warning items were worked through with little delay and the Town Meeting adjourned at 4:48 p.m.—one hour and seven minutes to transact the annual town business for another year. During the meeting, the two dozen Brunswick voters in attendance, sitting around an old iron woodstove sending its heat waves radiating toward the ceiling, adopted the 1992 Town Budget of $11,071. This was to pay for all town business during the year. Brunswick, some years ago, "threw up" all its remaining town roads—a vivid term indeed—meaning that under Vermont law they were no longer legally town roads at all, the lands reverting to the adjoining private landowners. This leaves the Vermont State Highway Department owning and responsible for the only current public highways in town, State Highways 102 and 105. The result is that since the 1960s, Brunswick (with a population in 1992 of 92 people) is spared any highway or road expenses of its own—or any road headaches. Frugality tends to be a way of life in the North Country.[1]

To complete the story of how Brunswick, Vermont, voted that first March Tuesday to govern and maintain itself for yet another year, we note that the same body of voters that afternoon, just previous to the Town Meeting, were also gaveled into being by the same Moderator Bates as constituting the Annual School District Meeting, a totally separate legal entity, and, having been duly Warned, proceeded to vote in the affirmative for a proposed budget to send its 29 students over the Connecticut River bridge to school in Stratford, New Hampshire (K–12). This "tuitioning out" of Brunswick children has been the annual practice ever since the small schoolhouse, with its under-the-same-roof outhouse in an attached back shed, was abandoned as a school and turned into the Brunswick Town House following World War II.

As noted above, the Brunswick expenses for the town business for 1992 were voted at just over $11,000. The School Budget was approved at the level asked for by the Brunswick School Board: $156,728. With the exception of a few thousand dollars in school financial grants from the state of Vermont, and various other small state grants for planning, every dollar for town and school expense was to be raised from taxes on real property and the 14,000 acres of land that constitute the Town of Brunswick. Ninety-five percent of those acres are forested, and one-half of those acres is owned by Brunswick's largest property owner: the Champion International Paper Company. The tax to be paid on those forested acres in 1992 was based on a valuation of $200 per acre; with the town tax rate for 1992 set at $2.25 per $100 valuation, each forested acre was thus to pay $4.50 into the town treasury.

This microcosm description of a town on the easternmost edge of Vermont's "Northeast Kingdom" (the poetic name coined by the late U.S. Senator George D. Aiken) could be extended to hundreds of similar small communities (in the precise New England understanding of "Town"[2]) throughout the 26-million-acre Northern Forest region. Many, of course, are larger, if not in area, certainly in population, and in the size and complexity of their government. Brunswick, as noted, has no school of its own, no town roads, depends on the Essex County Sheriff for police work plus a rare appearance of a Vermont State Trooper passing through, and is served by a fire department and a rescue squad—both voluntary—from across the Connecticut River in neighboring New Hampshire. Some Northern Forest communities are cities—Millinocket, Maine; Berlin, New Hampshire; Plattsburg, New York—but towns such as Brunswick are very typical.

There are places in the Northern Forest with no local government at all. Large parts of Maine and, to some extent, northern New Hampshire and northeastern Vermont are "Unorganized Towns" or territories with very few, or no, people and are run from the regional, county, or state level. It is beyond the scope of this chapter to discuss at length these areas and the Northern Forest future, but constituting as they do such a large and important part of the Northern Forest's New England section, these unorganized areas will be very important to the future, a fact already well recognized by state governments, timber and paper companies, and environmentalists.[3]

As major changes come to the Northern Forest they will be felt first by the residents of the small communities of the region. And these communities must not be ignored, for they are just as real and, in many ways, more enduring than other human constructs in this vast region still dominated by nature. The towns and their governments are here and will remain. They will not go away, and stubbornly, they will not die.

Consciousness of this endurance is felt more and more by local people. They have seen rapid change in many ways, one of them being, in the last decade or so, a disappearance of the familiar corporation names they took for granted, feeling perhaps they would always be the "Payer" name on their paychecks—names such as St. Regis Paper, Groveton Papers (always with the final 's' in the corporate usage), Great Northern Paper, The Brown Company. They are gone forever. What remains are the towns: Whytopitlock, Molunkus, Bingham, Skowhegan, Rangely, and Eustis; Errol, Berlin, and Stratford; Lemington, Brownington, Hardwick, and Enosburg Falls; Elizabethtown ("E-Town"), Malone, Tupper Lake, Lowville, and Deferiet. These and all the others across this great stretch of rural and wild area may have constant economic troubles, they may be hard-put to find jobs for their workers and tax money to finance their minuscule governments, and they could well be criticized on "efficiency of governing" grounds; and, if through magical powers and a time-machine we were to have the opportunity to settle North America all over again, we might not have "Towns" in their present form at all. But they are here, and they will endure.

Brunswick, Vermont, and the great majority of all the other towns in the four-state region were in existence long before any of the corporations were even conceived, and they have outlasted many of those same companies; many of the towns celebrated bicentennials in the 1970s, along with the anniversary of the nation's founding. Vermont's Secretary of State's Office notes that no organized town in that state has lost its organized status or charter in recent times, with a single exception (Glastenbury, in the southwest corner of Vermont, and not within the Northern Forest region). Maine's State Secretary's Office cites the same kind of stability there: What are organized towns, historically, tend to remain that way, and very few of the unorganized districts, or "Plantations," move toward formal "Organized Town" status.[4]

It is well known that forest industry, including pulp and paper manufacture, is the backbone of employment throughout the Northern Forest. The majority of the workers in that industry live in the small towns and cities. Here are the houses, the small part-time farms, the ever-present woodpiles, pickup trucks parked in the dooryards, along with snow machines and, lately, ATVs (all-terrain vehicles); here are the yellow school buses, or, in small communities, school vans, creeping along well before dawn light, and well after dark, in December and early January, carrying kids to the centralized, modern schools over distances that would be astounding to school-age families in the New York City or Boston area. Here, in short, is "home," with all the meanings and values that name brings. Regional and national environmental groups, corporations, land developers, second home seekers, all, at some point or other, will have to reckon with the existence of the towns, even if it is only a far-off nonresident landowner receiving the annual property tax bill.

Admittedly, it is not always "efficient" to have these towns throughout the Northern Forest region. The major paper companies, for example, usually pay their property taxes to the organized towns promptly, and try to maintain good relations with local officials via their regional forestry staffs. This is because not only are they apt to be, as in Brunswick's case, the largest landowners by far, but their forestry and mill workers are citizens of the towns themselves, and, in fact, are quite likely to hold town office at some time or other during their stay.

Yet, understandably, it is in the unorganized areas of northern New England that major landowners receive far easier tax treatment. In Essex County, Vermont, for example, timberlands in the six Essex County "Unorganized Towns and Gores," an area of well over 100,000 acres, pay only about 75 cents per acre per year net tax, far less than in organized towns. In northern New Hampshire (Coos County), woodland taxes in unorganized areas vary by local needs, some areas with no residents paying no taxes at all in some years. In a huge, worldwide emerging "Global Economy," wouldn't it be more "efficient" to allow the small towns to die?

U.S. federal government policies, indeed, at times have seemed to agree with that position. The decline of small rural communities in both the U.S. and Canada has been well documented. As one recent writer described:

> In America, parts of the Great Plains and the rural South come to mind. These are places where cultures once flourished, that are now being depopulated to the verge of extinction. At a time when capitalism and democracy are seen as triumphant worldwide, for a whole region of America to slide into a depopulated environmental wasteland represents an unnerving failure, at least of capitalism, perhaps in its betrayal of citizens—of democracy itself.[5]

Those familiar with Northern Forest issues in recent years will immediately note a contradiction here. Unlike the Appalachian and Midwestern coal mining regions that the quoted writer describes, and unlike the U.S. and Canadian prairie areas, to speak of "depopulation" in the rural areas of the Northeast is to alter markedly the description of the menace of change that ostensibly led to the Northern Forest Lands Study and the subsequent Governor's Task Force recommendation to create the present Northern Forest Lands Council. The problem in the 1980s was the land boom—

in the words of the Council, "new heights of second-home development, fragmentation of the land base, and other development pressures on these traditional timberlands."[6]

But in the early 1990s, as the Reagan-Bush era begins to recede and the Clinton Presidency begins, there is a picture in many Northern Forest communities of quite the opposite: no people, no jobs, and local tax receipts falling far short of local needs, especially for schools. The deep and stubborn recession, the loss of North Country jobs, the changing nature and worldwide "rationalization" of the paper industry, automation in the woods and mills, and, especially in Vermont (but also in some Maine, New Hampshire, and local New York dairy farm areas), the bleak outlook for farmers are all impacts that are leaving a deep sense of unease about the future for full-time residents of the Northern Forest who must earn their livings there.

To take just one of the problems cited above, that of the paper industry, note excerpts from a newspaper article from a daily paper whose "news beat," it could be said, is almost in the dead center of the whole Northern Forest paper industry's location: Berlin, New Hampshire.

NELSON HOISTS RED FLAG IN GORHAM

It is too soon to tell if James River will still be making paper in the North Country in another five years, James River Operations Manager David Nelson told the (Gorham, N.H.) Board of Selectmen Monday.

Paper is selling at its lowest prices since the recession began in 1989, Nelson said . . .

James River is not alone in its struggle to survive . . . one in three of the mills that were operating in 1970 are idle today . . .

An International Paper [Company] mill that shut down recently was losing $1 million a month and faced costly environmental problems, says Nelson, adding that James River's losses in Berlin-Gorham are on the same order of magnitude . . .

Strides in technology over the past 10 years have flooded the market with communication grade paper . . . James River's older paper machines cannot compete with . . . modern equipment that is capable of producing five to ten times more paper with the same amount of labor . . .

When the most recent round of layoffs are completed next fall [1993] James River's local work force will total 1,250 . . . that number will be reduced by another 150 through attrition over the next several years and continue to decline gradually after that . . .

Had the mill invested money in a modern paper machine, it would have added to the glut of paper already on the market, Selectman Glen Eastman observed. Nelson agreed, adding that if the mill did modernize to state of the art equipment, half of James River's employees would likely lose their jobs.

. . . We are relegated to our current path because we could not get the money [to modernize] said Nelson, adding that fortunately the current path is the friendliest path for the communities that are affected.

The layoffs that the North Country has seen over the past six months are minor compared to what would have happened if the plant were truly made modern [Nelson] says.[7]

What, then, is the problem facing the towns of the Northern Forest? Is it too rapid growth, with local values being overwhelmed, or is it decline and perhaps eventual death of the communities? What is happening, of course, is both conditions at once, at different locations around the region. While Lincoln, New Hampshire, debates effects of a proposed major expansion at the Loon Mountain Ski Resort, and nearby North Conway wrestles with severe auto traffic problems, Colebrook, 70 miles to the northwest, struggles with the loss of 60 jobs as the Bose high fidelity loudspeaker assembly plant there closes abruptly, moving operations to far distant parts of the continent. Another James River paper mill at Groveton, New Hampshire, just south of Colebrook, announces 80 more workers "furloughed"; and Mr. Nelson of James River, Berlin, gives his talk about the future of *his* mills to the Gorham Selectmen.

Some communities during this "downtime" of the early 1990s are using the breathing spell to do land-use planning and other forms of preparation for the land boom when (and if) it begins again. Others, perhaps understandably, are seeking any development they can find in an effort to keep up local economies. Still others are doing nothing, the residents perhaps picking up the cry of "property rights" advocates who have been using the economic slowdown as a stick with which to beat upon planners and environmentalists.

In this mix of uncertainty, vision or lack of vision, and different motivations, the towns and communities of the Northern Forest struggle on. The task of governing them seems to grow more difficult and time-consuming each year. Run largely by volunteers, the towns find their members of selectboards, planning commissions, tax assessors ("listers"), and school boards growing frustrated and tired. Resignations and turnovers increase. What used to be honored positions in some towns are seen now many times as thankless duties. State and federal mandates seem to increase endlessly. Local school boards, for example, often feel they are really local arms of state education departments. (A publication from the Vermont Secretary of State's Office, in fact, calls them just that.[8])

(One purely unscientific measure of how town-official profiles are changing, due in large part to the frustrations just cited, is an observation of who are filling these positions. In the past, new arrivals to an area who became permanent residents would have to wait years before being voted into an important town office. Now, however, posts such as Selectman/woman are being given gladly to "the new folks in town." Let *them* take the late-night telephone complaints about school bus service, a neighbor's dog chasing chickens or sheep, or perhaps a humped-up road culvert on a town road out back of beyond that has cleaned off "my exhaust system and prob'ly knocked the friggin front-end out of line too!")

Yet, despite all, the towns of the Northern Forest will continue. Somehow, volunteers come, offices are almost never left unfilled, and somebody attends to "town business." Another round of Annual Town Meetings ("March Meetings") approaches and with it the fierce pride of these citizens for the places in which they live. Here, then, is the challenge for those with an interest in, and a commitment to, the Northern Forest's future: How will the towns of the region be "dealt in" in the planning for that future?

No one should be terribly surprised if they deal themselves in, strongly. Local control will be fought for tenaciously, a factor recognized and understood by successful state and national political figures. Conservation policies and programs that also recognize this reality will have a much greater chance of succeeding. In the brand-new federal Forest Legacy Program, for example, in the four Northern Forest states that serve as pilot states (along with the state of Washington), some kind of approval from local communities will most likely be required before a Legacy parcel is finally concluded. Historically, also, Maine, New Hampshire, and Vermont have long had differing kinds (depending on which state) of town-level approval before new land acquisitions can be completed on the White and Green Mountain National Forests.

Some regional private-sector conservation groups are committed to the grass-roots, community-level approach, reaching out to local people throughout the region. The Vermont Natural Resources Council's "Northern Forest Project" has conducted over 35 "living room meetings" over a 2-year period, where 10 or 15 local people gather for an evening in a host home to talk over their visions and hopes for their town and region. The Society for the Protection of New Hampshire Forests has a "Voices of the Northern Communities" program conducted in conjunction with an area-wide regional planning and economic development group, the North Country Council. The "Voices" being heard are again those of the local people in their communities.

The concerns being raised by these local citizens cover a wide range, but major themes arise again and again: property tax burdens of farms and forest lands, local economic futures ("where will our kids find jobs?"), and clearcutting. Right or wrong, and despite the familiar explanations by foresters of how the practice can be beneficial when properly done, local people do not like large clearcuts in and around their towns, and, increasingly, are saying so. The "property rights" issue is often raised also, but, interestingly, among most local people (with the exception of a few usually well-known right-wing ideologues), a natural concern for one's property is increasingly balanced with the theme of working together for the common good in a time of uncertain futures.

And so, how will we, those of us deeply concerned with the Northern Forest, deal with the local people who live in the towns? Perhaps the best answer comes if we turn the question on itself and ask "How will *they* deal with us?" Can we work together for a common vision of the Northern Forest's 26 million acres, for common goals?

Ross Whaley, a forester, is the thoughtful and very capable President of the State University of New York, College of Environmental Science and Forestry, and was a New York member of the original Governor's Task Force on Northern Forest Lands. In some ideas he shared recently, Whaley noted some worldwide trends in political and social life toward localization. People, individuals, are playing increasing roles in decision making. The idea of smaller and more local governments is gaining favor. But just at this time, many have forgotten, or never learned, how to "do" local, grass-roots government. There are not many good role models of how to get locals involved directly in the task of governing with peaceable and good results. Switzerland comes to mind, and, for Whaley, perhaps some areas in Africa. Then also, there is New England and its Town Meeting tradition. "People in New England have a certain sense they are part of local government," Whaley says, "and in dealing with the Northern Forest issue, or anything else, to ignore this would be fatal."[9]

Here, following Whaley's thoughts, is the chance for the Northern Forest future to be guided, as we hopefully move toward a truly sustainable forest economy, without the rancor and bitterness that have been characterizing such efforts recently in other parts of the country. It is working with the towns, the grass roots, the local-government, decision-making models, that may be our best opportunity in the Northeast to avoid a polarizing "Northern Forest Chapter" in the apparently continuing tragic story of the Pacific Northwest spotted owl.

The Northern Forest of New England and New York is the only really large area in the mostly urban and suburban Northeast where nature is still the dominant force and influence. Situated as it is, midway between two of North America's largest urban concentrations in two countries—Montreal and the Boston–New York City–East Coast Megalopolis—the contrast between those areas and these 26 million wild acres is profound. A Vermont Natural Resources Council intern from Middlebury College, working for a summer on the Northern Forest as a student of resource economics, described her reactions to a stay in the region:

> Through my work, I discovered a wonderful place. An area where [Vermont's] largest farms are only one-half hour from the state's largest private timberlands. A place where the spruce–fir forest and the northern hardwood forest blend together. Lands crisscrossed by rivers and streams, dotted with lakes and ponds, and graced by unique wetland and highland areas. Spaces where bear cubs ramble playfully through the woods, and loons laugh while it rains.
>
> I saw a working land that gives us milk to drink, lumber to build with, and paper to write on. People and place come together in [the] Northern Forest Region to create a remarkable land with a natural vitality and a unique character.[10]

Sally Keefe's words put a human face on this wonderful land. "People and place coming together" is the best word-picture I can think of to describe the small cities, the mill towns, the rural communities, and hamlets, of the Northern Forest. For, once again, along with all the forest creatures, for the tenacious human dwellers of this place, here in the Towns of the Northern Forest Region is home. Here is where we will stay.

PART III

The Major Voices in the Debate: Timber Industry and the Environmental Community

Like many debates over management of natural resources in this country, what has kept the issue from being decided solely by government representatives has been the involvement of private citizens. In the case of the Northern Forest, the entire structure of the Northern Forest Lands Council, the primary advisory group in this region at the present time, is built on the recognition that many different viewpoints are held by the public and demand to be heard. There is no good term to describe the nongovernmental groups that play such a central role in this debate. They are often referred to simply as "special interest" groups, but this term carries the connotation that members of such groups are acting only for their own benefit. None of the authors in this section believe that to be true. All of them believe, with justification, that careful attention to their concerns and adoption of their visions are necessary for the betterment of everyone's future, not just their own.

The two voices that are primarily seen as defining the end positions in the debate are those of the timber industry and the environmental community. Yet neither of these perspectives is in itself homogeneous. We have asked two people who make their living from the forests in this region to present their perspectives on what the issues are and what is needed to achieve their vision of the future. Jonathan Wood is a forester with a relatively small timber-harvesting company in northeastern Vermont. His perspective is that of someone who believes strongly in the successes of private management of forest land, the importance of the Northern Forest in the future of the global economy, and the value of minimizing government interference in the culture and economy of the region. He has a long history as an articulate spokesman for private forestry interests, serving on local, state, and industry committees that focus on forestry issues.

Another perspective from the forest products industry is offered by Henry Swan. Unlike Jonathan Wood, Henry Swan specializes in timberland investment. His concern in this issue begins as a concern for the economic value of forest land as a source of investment. As such, he offers us a regional perspective on the economic status of the forest, using different indicators to assess the impact of past forestry practices and the need for forestry reform than does Jonathan Wood. He does not limit himself solely to an analysis of economic value, however, and also looks closely at what we must consider ecologically, culturally, and politically to achieve what he feels is the desirable future. As a member of the Governors' Task Force, which originally analyzed the extent of the forces that operate to change this region, he also offers a historical perspective on how far the debate has come since the mid-1980s. Clearly, other perspectives from the business sector exist, but these two have been among the most central in the debate so far and therefore offer us important insights into the opinions, concerns, and goals of the forest-related business community.

To some, the other end of the philosophical spectrum is held by the environmentalists. Yet, just as with the business sector, this group cannot be characterized by one position or spokesperson. Environmentalists themselves characterize their opinions as varying in the extent to which the existing political and economic structure of society must be transformed. Some groups, often characterized as "mainstream environmentalists," believe that solutions to environmental and economic problems can, for the most part, be achieved through the policy arena. Emily Bateson, Land Project Director for the Conservation Law Foundation in Boston, makes an exemplary contribution to this section from this perspective. The Conservation Law Foundation has worked extensively on the Northern Forest issue, primarily in Maine, with particular regard to the use of the existing process of hydropower relicensing on public rivers to effect changes in corporate policies. As such, it creates a link between public rights, such as access to and quality of public waterways, and private actions. Bateson's contribution clearly describes this perspective and incorporates it into her larger vision of a region that actively seeks to develop an ecologically and economically sustainable society.

Others within the environmental community have focused less on single issues and far more on long-term goals of how society can live while promoting healthy, intact ecosystems. Often called "radical environmentalists," their views on specific issues are often not that dissimilar from other environmentalists. They have, however, traditionally devoted more attention than others to questioning the social roots of the perceived problems and have questioned the wisdom of perpetuating processes that reinforce a destructive relationship with the environment. Jamie Sayen, editor of the *Northern Forest Forum*, is an ideal spokesman for this perspective. Sayen has a long history of involvement in the Northern Forest issue, and has a reputation, even among those who hold opposing views, as being an energetic and committed participant in the dialog. His chapter lays out a long-term vision of how society in this region needs to be transformed in order to achieve a future that guarantees a place for people as well as the rest of nature.

A common theme that runs through all of these chapters is that of sustainability, the idea that the needs of the present must be met without jeopardizing the ability of future generations to meet their own needs. Few people would argue that this trade-off between present and future generations does not exist or need not be considered. Indeed, each of these authors phrases his or her suggestions in terms of how and why they help the people of the Northern Forest region achieve sustainability. Where they differ lies more in what they view the constraints to change to be. These constraints are often phrased as "rights": conditions or agreements that provide the framework around which all policies must be shaped. Each of these authors also offer a perspective on rights, both private and public.

Together, the chapters in this section offer a broad range of perspectives on the concerns and visions of private citizens on the future of the Northern Forest. It is interesting to note how much common ground exists among these authors and where exactly the points of disagreement lie. Resolution of this issue depends on careful attention to their views and forging solutions that incorporate the essential elements of these visions.

9

Sustaining Our Forest—Crafting Our Future

Emily M. Bateson

Think Globally but Act . . .

"Save the rainforests." "Protect biodiversity." "Promote sustainability in developing countries." These are some of the new catch phrases of a burgeoning international debate, and for good reason. Truly magnificent failures by the World Bank and other similar organizations have led to the realization that we cannot promote economic development schemes that erode the very natural resource base upon which a society—and the world—depends. Horror stories emerging from the former Soviet bloc also illustrate that we cannot pollute at rates greater than the surrounding environment can absorb, or use up resources faster than the Earth can replenish. Tales from the Arctic to Zimbabwe depict the poignant fact that we cannot destroy habitat and species without risking our natural resource base and with it our cultural heritage and economic future.

This concept of the environment and economy being closely linked and complementary—rather than antagonistic—has led to the pursuit of "sustainable development" throughout the globe. The United Nations studied the problem for three years before publishing the memorable 400-page report, *Our Common Future*, in 1987. The Worldwatch Institute publishes a detailed study, and international bestseller, each year chronicling our successes and failures on the road to a sustainable future.

Then why is this dialog so absent from the Northern Forest debate? No matter how many recycling bumper stickers we buy, or pounds of Rain Forest Crunch we eat, the fact remains that we do not have—and do not promote—sustainable development right here at home in the Northern Forest. As we ponder the future of this region, with multivolume studies, multistate councils, and multiyear funding, the debate about the future of the vast Northern Forest region remains singularly one-dimensional: What's good for the forest products industry must be good for the region.

Alas, the world was never as simple as we thought when that famous reasoning was first used to shore up General Motors. In truth, it appears doubtful that propping up the existing forest products industry will save the multiple resource values of the region. Only when we break the current political gridlock and redefine the Northern Forest issue in terms of long-term sustainability—environmental, economic, and community—will we be able to grapple with the full range of pertinent issues and begin to craft an integrated plan toward a sustainable future.

What Is Sustainability?

The best known definition of sustainability was given by the World Commission on Environment and Development: "Humanity has the ability to make development sustainable—to ensure that it meets the needs of the present without compromising the ability of future generations to meet their own needs."[1] The impact of humans on the global environment has doubled under virtually all indices in the last 20 years alone, changing our historical perception that the two limits in an economic system are labor and capital. Instead, it has become increasingly clear that the true limits are the natural "sources"[2] (it does not matter how many fishing boats there are if the ocean holds fewer fish), and natural "sinks" (the capacity for our globe to absorb CFCs [chlorofluorocarbons] or nuclear waste is exactly zero).[3] We simply cannot afford to promote products, industries, or activities that are beyond the limits of our sources and sinks. Natural capital, in other words, has become a scarce and valuable commodity. And the problem goes well beyond a handful of troublesome toxics. With the current scale of industrial production, "even normally non-toxic emissions, like carbon dioxide, have become a serious threat to the global ecosystem."[4]

Periodically, we overshoot the natural limits on an ecological system but then attempt to back up and bring it back in balance—such as when we buy back fishing boats to protect a depleted fishery. The danger of this approach, however, is that if we wait too long to try and correct the problem, our ability to back up disappears and instead the ecological system collapses. In other words, we are flirting with a high-stakes game of global roulette—always hoping that ecosystems will rebound rather than plummet beyond salvage.

Faced with this daunting prospect, a new paradigm is steadily emerging. If environmental degradation is not to become irreversible, fundamental changes in the way goods are produced, used, and disposed of are unavoidable. A sustainable society will have to give greater emphasis to conservation and efficiency, rely more on renewable energy, and extract nominally renewable resources only to the degree that they can regenerate themselves. It will need to minimize waste, maximize reuse and recycling, avoid the use of hazardous materials, and preserve biodiversity. It will need to develop more environmentally benign production technologies, and to design products to be more durable and repairable.[5]

How are we going to achieve these lofty goals? Noted World Bank economist Herman Daly has proposed some operational principles of sustainable development.[6] First, the harvest rate of renewable resources should equal their regeneration rates. Second, waste emission rates should equal the natural assimilative capacities of the ecosystems into which the wastes are emitted. Third, the depletion rate for nonrenewable resources should be limited to the rate of creation of renewable substitutes by making sure that a suitable portion of the income gets invested in developing a renewable substitute in a timely manner. Two additional principles of sustainability can be added: Critical resources should be identified and protected from harmful future development, and the quality of life for people, their communities, and society as a whole must be enhanced.

Northern New England and upstate New York have not always been known together as the "Northern Forest," yet traditionally they have been characterized by similar qualities of hard work, thriftiness, and resourcefulness perhaps best summed up in the Yankee maxim: "Use it up, wear it out; make do, do without." Applying such old-fashioned common sense would be a good start toward the development of principles of sustainability for our own region. We should:

1. Develop an industrial system that depletes our natural resource base to the smallest extent possible and creates no waste, and an economic framework that supports that goal.

2. Protect our public and private land base and its ecological integrity through concerted and coordinated action from the local to the national levels.

3. Enhance local communities through environmental and economic literacy; diverse and community-driven industry; educational, cultural, and economic opportunities; and regional and national assistance with these community goals.

Sustainababble: Why Should We Care?

For a long time, my computer spelling-checker corrected "sustainable" to "sustainababble," and I confess that after a while I was inclined to agree. "Sustainability" and "sustainable development" are hurled into environmental debates with increasing frequency, often with little clear meaning and no concrete examples. The confusion is widespread. In one focus group in Vermont, for example, the group wrestled with the term "sustainable development," and finally decided that it might mean extremely durable housing.[7]

Although the ambiguity of these new terms has led some to discard them, I believe they provide an important new framework for the Northern Forest, a framework that rejects the notion that the economy and the environment are loathsome enemies and that banishes the myth that environmental protection leads to job loss and economic ruin. It is our challenge and our imperative to flesh out our new sustainability paradigm with specific goals and policies at the national, regional, state, and local levels. The Northern Forest debate provides a perfect opportunity to begin this critical task.

However, the current debate over the future of the Northern Forest region was inspired by the development boom of the mid-1980s that threatened to fragment large landholdings at the expense of public values and regional identity. Thus, "reinforcing the traditional pattern of land ownership" became a battlecry acceptable to the forest products industry, and remains the centerpiece of the four-state Northern Forest Lands Council (NFLC) mission statement today.

The infamous "Nash Stream" land sale that helped spark the debate—involving corporate raiders, land speculators, and overpriced state land buybacks—was now more than 5 years ago. Unfortunately, the Northern Forest debate remains a hostage to that narrow timber bias, particularly within the NFLC where the Maine representatives have threatened on more than one occasion to pull out of the process that will culminate with recommendations to Congress in 1994. Tragically, a major opportunity for a meaningful discussion on regional sustainability is slipping away.

The American Revolution was not really just about a tax on tea, nor is the Northern Forest debate just about subdivisions. The forest products industry is an important piece of our past and of our future, but the simple fact is that it is not the only piece. It is time for us in the region to expand the debate, to consider the bold and creative integrated strategies for the region left languishing in dark corners by the council. Otherwise, the spotlight of interest focused on this vast region ever so briefly will be extinguished, and we will have lost the political momentum, the public support, and the financial backing to craft a more sensible and sustainable future.

Crafting a Sustainable Northern Forest: Ten Basic Steps

Sustainability does not have one monolithic or static endpoint. There is no brass ring. Instead, sustainability represents an ambitious concept for reordering our industrial, economic, and social systems in order to live within our environmental means. My own framework for a debate on achieving sustainability in the Northern Forest follows. These and other sustainability issues must be pursued, refined, and implemented from the living rooms and town halls of the region to the halls and Houses of Congress. Only then will we define and reach toward a sustainable Northern Forest.

1. Promote eco-efficiency in existing industries. The ultimate driver of the global environmental crisis is industrialization, which means significant, industrial change will be unavoidable if society is to eliminate the root causes of environmental damage.[8] Emerging trends toward "corporate environmentalism" must be actively encouraged, including pollution prevention, energy efficiency, and recycling.

> Initially, business had a hard time taking environmentalism seriously, and saw the philosophy underpinning it as a passive, regressive, antigrowth and antitechnology—an attitude that made genuine action on environmental issues almost impossible . . . [whereas the emerging "green corporation"] takes the long-term view and addresses environmental issues by attacking their root causes. This new outlook . . . is founded on the recognition that environmentalism can be compatible with good business and is essential for business survival.[9]

In particular, in the Northern Forest region we must aggressively promote the development of "industrial ecology." The purpose of this emerging concept is to replace the linear "extract and dump" industrial pattern with a continuous cyclic flow of materials that learns from and mimics natural processes, using less energy, fewer material inputs, and fewer waste outputs.[10] This will have the result of transforming the current practice of putting out environmental brushfires as they arise to that of one that deals with the underlying causes of environmental degradation, thus precluding constant "environmental surprises."[11]

The Business Council for Sustainable Development, an international organization comprised of 60 of the world's largest corporations, has embraced the concept of "eco-efficiency." They believe that companies have an obligation to look at the life-cycle effects of their products and find ways to prevent pollution and strive for more efficient use and recovery of resources.[12] Remarkable results are possible even without exotic new technologies. The Fox River Textile Mill in Iowa, for example, recently increased productivity by 50% while reducing waste by 30% and energy use by 50%.[13]

Ideally, the result of industrial ecology efforts will be the creation of radically new manufacturing systems that have no waste. Any waste produced would become inputs for another production process. Industrial ecology thus provides the sorely needed comprehensive framework for industrial reform, resulting in the second industrial revolution. Without such a framework, the region may long continue its plodding "band-aid" approach, perpetuating the clash between environment and economy, and further eroding our economic and ecological capital.

In addition to cooperative efforts and incentives for industrial ecology, strengthened regulations and green taxes, such as the successful tax on CFC emissions, will be necessary to accelerate environmental progress and spur on recalcitrant industrial sectors. The recent attempt in Maine to loosen dioxin emission standards was not an advance in the direction of sustainability.

2. *Maximize energy efficiency in forest products and other industries.* Once billed as "informational terrorism" by the utilities, energy efficiency has caught hold in New England, and our region now leads the nation in aggressive, utility-funded efficiency programs that provide a powerful spur to regional economic health by lowering energy costs and creating the margin of environmental improvement necessary for economic development.[14] Maximizing energy efficiency through utility investment and industry innovation is essential in order to:

a. Help existing Northern Forest industries compete and expand. The $500 million invested by New England utilities in businesses and industries has resulted in lower energy bills, increased competitiveness, and expanded industrial production.

b. Create new jobs. Energy efficiency programs are labor-intensive, providing high-quality jobs in equipment manufacturing, construction, installation, monitoring, research, and development. These programs could create more than 30,000 jobs in New England alone over the next decade.

c. Reduce energy costs. Currently planned investments in energy efficiency by utilities in the 1990s alone will reduce electric bills in New England by approximately $6.7 billion. An expanded efficiency program could save New England $14 billion over the next two decades. Lower energy bills will increase the global competitiveness of Northern Forest businesses and the vitality of the regional economy.

d. Reduce environmental compliance costs. Much of New England and New York is in serious violation of federal clean air standards. By reducing energy use and the pollution it creates, efficiency programs already planned will save New England up to $1.6 billion in Clean Air Act compliance costs by eliminating unnecessary air emissions.

e. Provide benefits to the community. The cost savings associated with efficiency programs, coupled with utilities' targeted programs toward low-income housing, will be a welcome relief to ratepayers in hard economic times.[15]

3. *Promote use of secondary versus virgin materials.* The forest products industry and other industrial sectors must move rapidly toward recycling and source reduction for both environmental and economic reasons. This is one of the most basic tenets of a sustainable society. In addition, instead of burying and burning potentially valuable materials, the Northeast should exploit this resource to attract new manufacturing industries, create jobs, and stimulate regional economic growth.[16]

There is a clear trend in both the marketplace and the regulatory arena toward increased recycling.[17] Increased costs for waste disposal are creating new supplies of secondary materials, and new demand for recycled goods is creating new market opportunities. In addition to easing pressure on the region's overflowing landfills and communities' strapped wallets, recycled paper processes generally use only 10 to 15% of the bleaching necessary in the virgin process,[18] use up to 60% less energy,[19] have lower transportation costs, and have far lower capital costs for mill expansion.[20] In addition, shifting to use of secondary materials will capture emerging new markets[21] and hold onto existing market share.

The Bowater newspaper recycling plant is a good example. In 1991, Bowater, Inc., acquired a majority ownership of Great Northern Paper in East Millinocket, Maine—the only newspaper producer in New England. With the assistance of the Finance Authority of Maine (FAME), Bowater started construction of a major new de-inking facility and recycled paper operation. Since its completion in 1993, the plant has the capacity to accept 140,000 tons per year of old newspapers, telephone directories, and magazines. Designed to accept waste from roughly 6 million people (or a population five times that of Maine) the Bowater plant will be a major benefit to the communities of the region by stabilizing newspaper prices and reducing tipping fees for solid waste. Over 300 jobs were created during construction of the facility, and up to 30 permanent employees are needed to run the new operation. In addition, many jobs will be maintained through the increase of collection, transportation, and processing of waste-paper. It has been estimated that roughly one job may be supported for every 465 tons of material collected and processed, meaning that Bowater may spawn approximately 300 new jobs in the Northeast.[22]

In addition, the facility may well save Bowater upward of $50 per ton of paper produced.[23] It will also allow Bowater to retain its major clients, including the *Boston Globe*, the *Wall Street Journal*, and *USA Today*, as they attempt to comply with minimum content agreements reached with their states. As Bowater Chairman A. P. Gammie noted at the ground-breaking ceremony: "And of paramount importance, [the new recycling mill] will secure the continued business of a number of customers who otherwise had threatened to leave."[24]

It is difficult to estimate to what extent Bowater's switch to secondary material source will decrease harvesting pressures on the Northern Forest. However, one recent analysis estimates that recycling a ton of old newspapers may save about 12 trees.[25] In theory, the Bowater plant will thus decrease demand for pulpwood to the tune of roughly 1,680,000 trees per year—a considerable environmental savings.

Paper and allied products recycling represents a growth industry, a value-added industry, and ultimately a necessary shift for the Northeast if we are to retain our existing domestic and international market shares. Located in a populated region that clearly excels at producing garbage, the Northern Forest paper industry may finally have a niche market that provides a competitive edge over the larger, newer mills of the South and Canada—if we seize the chance.

4. *Maximize value-added opportunities in all industries.* The North Country is increasingly operating as a Third World country, exporting its most valuable raw material to other regions for further processing. Active, sustained intervention will be required to turn this trend around.[26]

It is essential that we maximize the number of jobs created from each tree harvested by encouraging value-added manufacturing. Otherwise, export of raw logs to Canada and elsewhere will increasingly send jobs across the border. For example, at the height of the Pacific Northwest's uproar over the so-called "spotted owl problem," exports of raw logs from Oregon and Washington to Japan reached 24% of the timber cut.[27] With the value of American hardwood generally increasing on the international market, it is time that we give this issue serious scrutiny. Supplemental strategies should also be considered, such as incentives for long-term improvements to the quality of the wood grown and harvested—including more labor-intensive pruning and thinning—in order to promote long rotations of hardwoods and related opportunities for value-added manufacturing with hardwoods.

5. *Identify and promote new, diverse industries that are sustainable.* Despite the predominant focus in the current debate on the forest products industry, other employment opportunities are clearly on the rise. For instance, tourism now accounts for a higher payroll in Vermont than forest-based manufacturing, while region-wide, state general revenues from forest-based tourism exceed that from forest-based manufacturing.[28] On average, each thousand acres of forest land in the region supports 1.9 forest manufacturing jobs and 2.8 forest tourism jobs.[29]

In the Bowater mills in the Millinocket, Maine, area, mill employment has declined almost 50% just in the last decade, from more than 4,000 jobs in 1981 to just over 2,000 today—despite a relatively stable product output. On the other hand, management of the watershed to enhance recreation consistent with traditional back-country use would create significant economic benefits and diversify the economy. Even three to nine small sporting camps or canoe huts carefully spaced a day apart would enhance the current four-season use of the area, provide $1.74 million and $6.79 million in direct and indirect revenue, and create up to 180 new jobs. A marketable 2- to 3-hour whitewater rafting experience appropriately sited between Millinocket and East Millinocket would be similar to a site on the Ocoee River in Tennessee, which generates over $25 million annually.[30] However, new opportunities need not be tourism-based, as long as they meet the basic objectives for sustainability in the region.

6. *Protect core areas for ecological integrity.* The Northern Forest region is characterized by the last large stretches of undeveloped lands in the Northeast, and supports the numerous public and ecological values on which we all increasingly realize we depend: watershed protection, biodiversity, wildlife habitat, recreational opportunities, and water quality, to name just a few.

The many simultaneous uses of the Northern Forest have led to it being termed a "working forest." However, the legitimate question has been asked: *Is* the working forest working? Ultimately, I believe the answer is no. A predominant industrial use of the Northern Forest in our current economic system cannot be expected to single-handedly preserve ecological values to the extent necessary in a region so lacking in public land. It would either be politically impossible to implement, or economically unfair to impose. Instead, it is essential that environmental sustainability in the Northern Forest start with the establishment of large, protected, ecological reserves, connected by protected corridors. Only in this way can we begin to protect this vast and vulnerable land base for economic security and durability for the years to come.

Perhaps the most promising dialog to emerge from the NFLC process to date is the recent discussion on biological diversity. Although dodged for several years by the Council and the Governors' Task Force that preceded it, the pivotal topic of biological diversity has finally been raised, and appears to be a permanent part of the debate. In a real watershed for the Northern Forest debate, the NFLC hosted a Biological Resources Diversity Forum in December 1992. This important gathering reflected remarkable agreement among the nine invited experts concerning the need to create a system of buffered, connected ecological reserves to protect the ecological integrity of the region. At the forum, wildlife biologist Malcolm Hunter of the University of Maine termed the biological diversity in the region as "poor and declining," while Sharon Haines of International Paper agreed that "we are going to have to have preserved land. There's no question about that."[31]

A subsequent draft briefing paper on creating ecological reserves, prepared for the Council by Hunter and Haines, explains that the "most important goal of an ecological reserve system would be to maintain the region's biological diversity."[32] Since most of the region's biodiversity resides with invertebrates and other small species "about which we know almost nothing," the authors conclude that "this coarse-filter approach is essential," and would "significantly help stem the tide of species that are declining to the point where they require special protection."

It seems increasingly clear that the Northern Forest region should be characterized by a series of substantial, undisturbed core areas connected by corridors. Emphasis should be on "representing all ecosystems, maintaining viable populations of all native species, maintaining ecological and evolutionary processes, and being responsive to change."[33] These reserves must be part of a larger, integrated approach to private and public ecosystem protection, given the emerging consensus that such key reserves will not, in and of themselves, safeguard our natural heritage and sustainable future.

This pivotal protection effort should begin immediately by taking advantage of major, current opportunities. A number of examples spring to mind. First, there are currently an estimated 3 to 5 million acres of land on the market in the Northern Forest. These "willing seller" purchases are an immediate opportunity to provide a solid cornerstone to the ecological reserve effort. Second, the hydropower relicensing process, discussed in greater detail later, presents a unique opportunity to focus on river and watershed protection throughout the region, trading use of public waters by private companies for long-term habitat and corridor protection. Third, reform and expansion of two federal programs could dramatically help this pivotal core areas effort: the Land and Water Conservation Fund (currently biased toward funding western states with significant existing federal land base), and the Forest Legacy Program (a new program for easements and purchase of forest land).

7. *Manage private land for sustainability.* Even if society undertakes radical efforts to curb its impacts on remaining natural areas, most of the world's landscapes will continue to be dominated by human beings, and most will fall outside the scope of strictly protected reserves. Seminatural areas—such as second-growth forests, waters whose fish are intensively harvested, and rangelands grazed by livestock—prevail around the globe. Unless society can learn to tolerate and maintain wildness in these civilized landscapes, biodiversity has a bleak future.[34]

The reluctance of the NFLC to examine forest practices was, of course, a highly political decision, and perhaps necessary to hold together the four-state coalition. Unfortunately, it also sounded a death knell for logical discourse on the future of the region. By narrowing the debate, policymakers essentially asked us to take on faith that large blocks of land—no matter how they are managed—will provide the myriad of public values we seek to protect.

Despite the assumption in the Council's ecological reserves briefing paper that any reserves created in the region will be nestled in a landscape of "seminatural ecosystems," the people and the wildlife of the region in truth have no assurances that such will be the case. Although it is clear that some forest management activities can complement conservation goals, others do not. Can we save the forest but not the trees?

The NFLC's own briefing paper on ecological reserves concludes that reserves, by themselves, will not be sufficient to maintain biological diversity and notes that species of vertebrates and plants known to be rare would still require special protection efforts—the so-called fine-filter approach.[35] It is thus both logical and essential that surrounding private lands must be managed to provide corridors, buffers, and extended habitat systems for these central cores.

The remarkable ability of some notable ecosystems, such as the tropical rainforests, to contribute to global health and biodiversity appears to have led to a diminished appreciation of preserving biodiversity in our own backyards. However, it would be tragic to preserve a few large, famous habitats around the world while paving over everything else that does not measure up. Biodiversity should be protected at global, regional, and even local scales. Similarly, it would be tragic to expect a few habitat islands in the Northern Forest to provide every ecological need while we go about our (consumptive) business as usual. Instead, we must develop an integrated approach, with sensitive private land management and protected corridors complementing the fully protected ecological core areas.

We must begin by setting standards for sustainable forestry for different habitat types, at both the community and landscape levels, followed by a corresponding shift in economic incentives and disincentives to achieve our goals. State and federal grant programs, subsidies, and various tax incentive programs must all reward achievement of standards and guidelines for sustainability. The public must no longer be asked to subsidize forest practices that degrade the environment and overlook the long-term ecological and economic health of the region and its local communities.

8. Enhance the sustainability of public land and public property rights: protecting the public from the private. There has been considerable heated debate in the region about private property rights, including the emergence of a "wise-use" movement whose members share the common objective of encouraging unrestricted private economic exploitation of and access to this country's natural resources by systematically dismantling state and federal environmental laws. However, the quieter skirmish is over *public* property rights, and whether we will protect them or continue the quiet chiseling away of public rights and values that is slowly privatizing the region.

More specifically, in New England, the public's rivers have been harnessed to provide power and profits for private industry and utilities, and this year alone, 61 of those dams are up for new 30- to 50-year licenses. At stake are some of New England's major rivers, including the Androscoggin, Penobscot, Kennebec, and Deerfield. The initial licenses were issue for 30 to 50 years to allow investors to recoup their investment, but policymakers crafting federal water policy that culminated in the 1920 Federal Power Act carefully avoided an automatic right of renewal that might cede these public resources to private interests forever. As explained with considerable foresight by President Theodore Roosevelt:

> My position has simply been that . . . it shall not be a grant in perpetuity . . . . Put in a provision that will enable our children at the end of a certain specified period to say what shall be done with that great material value which is of use to the grantee only because the people as a whole allow him to use it. . . . [M]ake the grant for a fixed period, so that as conditions change . . . our children—the nation of the future—shall have the right to determine the conditions under which that privilege shall be enjoyed.[36]

Since the last licenses were issued in the 1950s and 1960s, the nation has undergone a transformation in environmental law and policy. Hydropower operator applicants in the 1990s are subject to numerous environmental laws to ensure that the dams are operated in a manner that minimizes intrusion on natural ecosystems and finds a balance between the value of the electricity generated and the public values and uses of the river.

Most notably, amendments to the Federal Power Act in 1986 require that the Federal Energy Regulatory Commission (FERC) "give equal consideration to the purposes of energy conservation, the protection, mitigation of damage to, and enhancement of fish and wildlife (including related spawning grounds and habitat), the protection of recreational opportunities, and the preservation of other aspects of environmental quality."[37] Coupled with other major environmental statutes, including the National Environmental Policy Act and the Endangered Species Act, these amendments make hydropower relicensing a far more balanced process in the 1990s than it was before. Or do they?

Unfortunately, indications from FERC and the hydropower companies are that Roosevelt's principles have been forgotten. We now find ourselves fighting to ensure that public rights established at the beginning of the century will be respected as the century comes to an end. The companies apparently consider themselves entitled to free, almost unconditional use of hydropower resources, and FERC seems inclined to agree. For example, Bowater, Inc., obtains between 60 and 80% of the power for its two paper mills from five generating dams and 12 storage dams on the West Branch of the Penobscot River in Maine. Bowater is also the largest landowner in the entire state of Maine, with more than 2.1 million acres of prime watershed, wildlife habitat, and recreational land. Bowater has estimated the value of the power generated by its dams at more than $270 million over the next 30 years. The Conservation Law Foundation (CLF), using a different formula, estimates the value in excess of $650 million. Either way, Bowater is asking for the exclusive right to hundreds of millions of dollars of power. The source of that power, the Penobscot River, is a public resource. What is the public getting in return? So far, not much.

These riverways and watersheds are important biological and geological features of our Northern Forests. As we work toward wise management of private lands, we must no longer neglect wise use of these public riverways. Otherwise, our generation will not get another chance. Northern Forest advocates should join the hydropower relicensing fray by advocating in each relicensing proceeding throughout the region[38]:

a. Significant land and water protection in return for harnessing public water for private profit.

b. Substantial mitigation funds for river conservation, watershed protection, and acquisition by setting aside a percentage of gross power revenues of utilities. A recent national survey discovered that 88% of the 1,009 polled favored requiring utilities to set aside a percentage of their hydropower profits for this purpose.

c. Relicensing decisions based on comprehensive river plans and environmental analysis, rather than a piecemeal, dam-by-dam approach.

d. Protection of riparian and watershed lands by having utilities donate such lands into public ownership or grant permanent, protective easements on such lands.

e. Ensure free, public access to the rivers in the vicinity of the dams.

f. Ensure maximum energy efficiency before a license is granted, and consider all other alternatives to relicensing.

g. Restore sufficient water flow to the rivers, and mandate upstream and downstream fish passage.

h. Promote rigorous oversight of FERC, and consider removing hydropower jurisdiction from FERC because of the agency's failure to balance environmental concerns with energy development. Regulatory authority could be assigned to an agency with water and natural resource management expertise, such as the U.S. Fish and Wildlife Service or the Environmental Protection Agency.

i. Shorten dam license periods in order to better reflect changing environmental conditions and public need.

This may well be the best leverage the public currently has to achieve better protection of our Northern Forest land and water resources at a reasonable fee, since we are already giving out such ecological largesse at no cost.

In addition to a focus on hydropower relicensing on public rivers, we must also focus on the management of public land. Many within the environmental community feel that some public acquisition of available land in the Northern Forest must be part of a long-term strategy for the region's future. Further, the land protection envisioned in an ecological reserve system may well be more logically managed by natural resource protection agencies such as the U.S. Fish and Wildlife Service and/or the National Park Service, perhaps in new partnerships with state agencies or regional entities. However, the track record so far of federal agencies, most notably the U.S. Forest Service, to manage public lands has not been impressive.

Do past failures of federal agencies indicate the futility of involving them in strategies for land management in the Northern Forest? I do not believe so. For example, radical statutory reform of the U.S. Forest Service would not be required to improve their success in managing healthy, sustainable forests. Virtually all of their failures, including below-cost timber sales, excessive logging at the expense of other values, disregard for wildlife protection, and excessive clearcutting, are specifically discouraged or prohibited by federal statute. The Forest Service has simply disregarded congressional mandate ever since a long series of Forest Service reform bills began with the passage of the Multiple Use–Sustained Yield Act of 1960. This is of particular concern in the East, where the small forests in our populated region are subject to Forest Service policies developed for the vast forest lands of the West, which too often deal inadequately with the special needs of the East.[39] There is no question that reform of the Forest Service, and careful review of other agencies (e.g., the Adirondack Park Agency) and acquisition within and beyond the borders of these public lands, are key elements of our land protection efforts.

To give just one example of how this reform can be brought about, a new planning process for national forests will begin soon. Because a portion of the White Mountain National Forest is within the Northern Forest, and the Green Mountain National Forest is a logical choice for public acquisition and management of land in the Northern Forest region of Vermont, public involvement in this planning process is crucial for the future of the Northern Forest. I believe that the public needs to emphasize four points with respect to the management of Forest Service lands.

The first is the protection of roadless areas. Take the case of the White Mountain National Forest (WMNF). Its management plan states that it provides a remarkable 70% of the primitive back-country recreation in the entire Northeast. The WMNF also supports over 300 species of wildlife, including a number of threatened and endangered species such as the peregrine falcon, bald eagle, and canada lynx. However, the 1986 Forest Plan crafted by the Forest Service increased timbering in the WMNF by 55% and proposed 162 miles of permanent new roads—contrary to growing consensus concerning the need to manage eastern national forests for the natural and ecosystem qualities rapidly disappearing in New England. In the next planning process, we must ensure that both of these forests preserve their roadless lands and ecological integrity, as the Green Mountains set out—on paper—to do in 1986:

> We believe that public land in New England is scarce and precious; our management philosophy reflects that belief. The Green Mountain National Forest should be managed to provide benefits that private land does not, and to maintain options and opportunities for the future. . . . this will become even more important as the land in New England becomes further subdivided and developed.[40]

The second point to emphasize is the protection of wildlife. The 1986 WMNF Management Plan's wildlife strategy was designed to correct a perceived deficiency in certain wildlife species and their habitats. Interestingly, those species are the ones that use early successional habitats that can conveniently be provided by timbering. The planners further divided the WMNF into a checkerboard of "moose-units" and proposed to maximize diversity in each unit. We must ensure that the next forest plans take a broader, regional view of the wildlife protection, and reject any attempts to maximize diversity through forest fragmentation and increase of the early successional species that already predominate in the developed landscape.[41]

The third point is reform in timber-cutting practices. The issue of extensive timber cutting on our national forests must be reconsidered, in accordance with both new national initiatives of the Clinton Administration and the changing needs of our region. In a region so deficient in public lands, it is time to more carefully manage those that we have. The 1986 WMNF Management Plan, for example, increased timbering levels from 31 million board feet (mmbf) to 38 mmbf in the first decade, although the proposed sales would result in a net loss to the federal treasury of $0.54 for every dollar spent, equivalent to $700,000 annually and $6.9 million for the first decade,[42] an amount that would buy a considerable amount of Northern Forest land.

The fourth point is the promotion of community stability. While timber harvesting contributed about one-quarter of local employment in the WMNF region three decades ago, this share had dropped to less than one-tenth by the late 1970s, primarily because of the growth in tourism and recreation-based employment. In addition, local timber industries are buying smaller percentage of WMNF timber. Although the WMNF Management Plan claimed that the below-cost timber program enhances the well being of local communities,[43] Forest Service policy cautions against such an assumption, "especially since increased dependency upon submarginal timber sales would seem to result in potentially greater community instability due to uncertainties over continuation of a relatively high level of federal funding to support a timber program with costs greater than revenues."[44]

The next forest planning process must more carefully consider whether impairing roadless and back-country areas on the forest with road building and timbering will degrade the very resources that have in large part propelled the area's burgeoning tourist and recreation industry.

9. *Sustaining local communities.* There is a legitimate presumption in the sustainability debate that development within our environmental means will provide a more healthy, stable foundation for local communities to flourish in the long term. This presumption is founded on several beliefs: (1) sustainable industries will soften the boom-and-bust industrial curve, with industrial diversity providing further stability to the regional economic base; (2) jobs in sustainable industry tend to be more labor-intensive, changing the current downward curve in timber employment, while development of sustainable technologies and sustainable products represents a

number of growth industries that will provide a real economic boon to our region; (3) environmental sustainability will protect the burgeoning tourist and recreation-based economy and provide more flexibility for future options; (4) resource protection is cheaper than subsequent "end-of-the-pipe" cleanup or long-term chronic health problems from environmental pollution; and (5) a shift toward eco-efficiency will help preserve the communities' natural resource base and quality of life, in particular slowing the unacceptable tendency to site major polluters and waste repositories in poorer rural communities for a fee.

However, crafting and enhancing a community also involves a number of tangible and not-so-tangible issues that must be a major focus of the dialog on sustainability in the Northern Forest, including:

a. Environmental and economic literacy. Communities must be informed of the issues so that they see the need to work toward economic and environmental sustainability. Without local commitment, these goals cannot be successfully achieved.

b. Educational and cultural opportunities. The greatest resource of the region is the people, and they must have sufficient education to take advantage of employment opportunities in their own communities.

c. Economic diversity and community-driven industry. Diversity of economic opportunities is plain common sense—don't put all your eggs in one basket if you can avoid it. In addition, community control of industrial opportunities puts the local residents at the higher end of the profit curve—it is better to be the boss than the low-wage worker—and keeps the profits in the community. Local control helps insulate the community against decisions made in the boardrooms of international corporations—decisions that can result in a major industry degrading, shortchanging, or leaving the community.

d. Industrial clusters. However, communities must guard against becoming too insular. It is well documented that community efforts are most successful when coordinated on a broader scale with regional assistance.[45] A clustering of similar industries in one area is often necessary to develop an adequate local advantage and network for a stable industrial base.[46]

10. Linking economic, ecological, and community sustainability goals in an integrated, innovative framework. The challenge is to adopt policies that make economic and ecological imperatives converge, and redirect market forces to achieve environmental goals. Since private businesses are by their nature focused on earning profits, it is up to governments to ensure that the most profitable investments are the ones that are environmentally sustainable. Policies ranging from well-crafted regulations to environmental taxes can be used to achieve these goals.[47]

Innovative corporations are beginning to think green, and the reason for this is also green: money. Traditionally, corporations internalized profits and externalized costs, making society pay for environmental degradation to our air, water, and land. Now that industry is being asked to cover some of that "cradle-to-grave" cost of doing business, pollution reduction and environmental protection have become cost-effective: It is cheaper to produce a substitute than pay a tax for producing CFCs, or to recycle than pay large tipping fees for solid waste.

Furthermore, it is increasingly clear that environmental regulation does not place an economic stranglehold on the state or region. Quite the contrary. For example, Professor Steve Meyer, of the Massachusetts Institute of Technology, found that states with strong environmental regulation actually grew faster economically than states with lax rules.[48] In particular, Meyer compared the regulatory climate of New Hampshire to the stricter environmental standards in Vermont and concluded that "New Hampshire's relatively lax environmental regulations may have aggravated the collapse of the state's real estate, development, and banking industries."[49]

In the Northern Forest, it is imperative that we (1) encourage sustainable practices by ensuring that the economic system maximizes the profit incentive for sustainable industry, and (2) create a new institutional framework in order to effectively link economic gain to achievement of sustainability objectives. It is imperative that we aggressively pursue a new framework that changes the economic playing field and promotes sustainable goods and services. We must provide economic incentives and tax relief to landowners, communities, and businesses that achieve specified sustainability objectives, and create a new profit structure that maximizes financial return for sustainable goods and practices. Our task in this area is immense.

However, energy efficiency in the region provides a compelling example of the need to create a clear profit motive for choosing the sustainable pathway. Five years ago, as CLF and others worked with utilities to sell efficiency instead of electricity, the profit motive was absent. Moreover, efficiency could actually result in lower sales between rate cases, with potential loss of profit margins. Since that time, New England's regulators have addressed these problems directly through a number of structural changes. For example, utilities are now entitled to earn a profit on their efficiency investments, based either on a percentage "share" of the total cost savings created or on a return on capital investment for direct efficiency. Massachusetts Electric, for example, projects that it will earn approximately $6.9 million in after-tax profit for successful implementation of its $55 million 1991 program. As the CEO of Massachusetts Electric's parent company noted, these regulatory changes "have turned conservation from being the most controversial expenditure I have to deal with to being the most profitable investment I make."[50]

Some economic changes can be made on a piecemeal basis, similar to the advances made in the region's energy efficiency programs in the past 5 years. However, many tax and incentive programs should be redirected in an integrated way, and targeted effectively at the landowners, communities, and industries that achieve articulated sustainability standards and guidelines. Rather than awarding the forest products industry further tax subsidies, as has been often proposed throughout the Northern Forest debate, any such economic incentives should be directly tied to achieving forest management and conservation goals. Such a quid pro quo program—referred to as an "elective forest conservation tax program"[51] in an analysis done for the NFLC—must be the cornerstone of any tax relief effort in the Northern Forest.

However, an economic sustainability package must also cast its net wider and encourage local communities to protect their environmental heritage and their economic well being. Although regulators must take the lead in creating a new economic structure, it is the local communities that must take advantage of that new framework and choose their own path toward a sustainable future. The current proposal in New York to provide tax breaks to communities that have a minimum percentage of preserved land is a fine example of a financial return on environmental protection. Such broader efforts that focus on redirecting a whole area in a comprehensive way toward an environmentally and economically sound future provide the most innovative and promising mechanisms for sustainability.

It may not be possible to restructure our global economic systems overnight, but there is no reason we cannot begin by recrafting our Northern Forest region. Our goals of industrial and economic reform, core area protection, sustainable private land management, and economic health of local communities are complex but complementary, challenging but achievable. Linking economic and environmental goals is critical; regional commitment to a bold new strategy is essential. In short, I propose the development of a zone within the Northern Forest that promotes a transformation toward environmental and economic stability.

The Northern Forest Enviroprise Zone

An enviroprise zone in the Northern Forest would involve five key elements. First, federal legislation would recognize the national importance of the Northern Forest and establish by statute a Northern Forest Enviroprise Zone. This zone would constitute a new federal–state partnership, with the federal government providing consistent and rational standards, guidelines, and timetables, as well as substantial technical, financial, and land acquisition assistance. The states would implement and manage the program through expansion of current agencies (such as the Land Use Regulation Commission in Maine and the Adirondack Park Agency) and ensure that any agency actions were consistent with state and local community goals. Communities would continue their current autonomy, with considerable new financial and technical assistance to aid their decision making and enhance economic opportunities.

Second, the statute would provide rough boundaries of core areas and corridors for the region to ensure that later state discussions did not lose track of regional landscape considerations, and provide funding for the states to hold hearings and finalize core area and corridor boundaries within a specified timetable. Also included would be criteria such as percentage of land that each state should zone and acquire as core areas, and specific guidelines for management. Review of federal and state programs, such as those for hydropower relicensing, for consistency with this objective would be funded. Monies would be provided to the states to inventory lands, hold public hearings, and finalize core areas and corridor boundaries. Funds would be provided for acquisition of these lands.

Third, the legislation would set out standards and guidelines for establishing carefully managed and adequate lands around core areas. The type of management and density of development would be specified. Monies would be provided for states to determine buffer zones and fine-tune standards to meet local needs. Management standards in buffer zones would be required; management standards in other private land (e.g., forestry, agricultural, and industrial) would be optional if they meet or exceed current state standards; however, substantial tax advantages and economic incentives would accrue if the guidelines were met.

Fourth, standards and guidelines would be set out for communities, and development of a local management plan would be recommended. Planning monies would be available, and economic grant funds would be available for implementation of approved community objectives.

Fifth, a special program would assist each state in reviewing its economic system and recommending substantial changes to encourage sustainability objectives. Grants would be available for industries to invest in new technologies that allow them to meet these objectives. An elective conservation tax package would be created that landowners, businesses, and communities would be eligible for if they met specified objectives—from forestry tax relief to estate tax relief for family owners in return for permanent protection.

It is difficult to so briefly summarize a complex regulatory mechanism, and this is by no means complete. However, it may be reassuring to note that the New Jersey Pinelands is governed in a similar way, in which the federal government provides financial and technical assistance, and up to 75% of the cost of core area acquisition, but leaves almost all regulatory and management responsibility with the state.[52] In Washington and Oregon, the Columbia River Gorge Act tackled the same issues of economic and community sustainability, allocating $40 million for land acquisition, $5 million for grants and loans for projects consistent with the area's scenic objectives, $2.8 million for transportation improvements, $20 million for interpretive and conference facilities, and completed a community economic opportunities study—all for an area of only 250,000 acres.[53]

Conclusion

The Northern Forest will have begun to invest in a sustainable future once we:

1. Protect substantial core ecological reserves surrounded by buffer areas where disturbance is minimal and connected by ecological corridors.

2. Surround these core areas with private land that is managed and developed only in a sustainable manner.

3. Redirect the economic system to promote environmental sustainability for industries, communities, and individuals.

4. Create an innovative framework for promoting and rewarding sustainability in a consistent, regional manner, modeled after programs such as the Columbia River Gorge and the federal enterprise zone program—an "enviroprise zone" for the Northern Forest.

We must think big, and we must act soon. The spotlight of attention currently focused on the Northern Forest region may well start to dim in 1994 with the disbanding of the Northern Forest Lands Council. It is not enough to worry about sustaining ecosystems halfway around the globe. We must get to work on sustaining our own Northern Forest.

10

A Sustainable Resource for a Sustainable Rural Economy

Jonathan Wood

In the cupped hands of the future lay other seeds, with fairer promise.
Donald Culross Peattie, *Flowering Earth*, 1939

My perspective on the Northern Forest issue is that of an industrial forester, but the vision I have for the future of this region is a result of my own values, which have been shaped by my upbringing, education, training, experiences, and hopes. I am concerned with the future of the Northern Forest because it is my home. It is where I live, work, recreate, and raise my family. I want for my children even better opportunities to enjoy the forest than I have. I believe that most of my views on this issue are shared by the forest products industry in general.

I am fortunate to have a career that allows me to spend a great deal of time in the forest. I choose to work in forestry for that very reason. Much of my free time is also spent engaged in experiences that connect me to the forest: hunting, fishing, camping, and skiing. My connection with and love for the forest is typical of many of the people who work and live here.

A quote from the Northern Forest Lands Study sets the framework for my vision:

> The special character of northern New England and New York, and the large forest tracts that now contribute to the economy of the region and the enjoyment of its people are the keystone of the vision for the future. The forests—including the air, water, soil, plants, and animals—must be healthy and available for all to enjoy.

A working landscape of forests interspersed with lakes, rivers, mountains, crop and pasture land will provide attractive visual contrast. The many attributes that create the special character of the region will be preserved, while certain of the economic, social, and environmental factors which affect people's lives must be improved.

The people of the area will play a central role in guiding the future conservation and management of the Northern Forest. Communities around the region will continue to be an important part of the working landscape. A modern forest industry will benefit from a steady supply of wood fiber. Tourism and other industries will add to a diverse economy. Development will occur in a planned manner, complementing existing settlements, and in harmony with the health and productivity of the region's forests. Citizens will benefit from good jobs, affordable housing and adequate health care. People will continue to have many opportunities to enjoy a wide variety of outdoor recreation such as fishing, boating, hiking, cross-country skiing, hunting, and snowmobiling. Large blocks of undeveloped land, connected by travel corridors, will provide these opportunities. People may pay a reasonable fee, but will otherwise be welcome to visit. Recreational use will exist along with forest management and will be compatible with forestry activities. Healthy diverse populations of all native plants and animal species will exist in a variety of ecosystems. Many opportunities will exist for people to view and harvest wildlife. Within the working landscape, there will be places that are protected from human disturbance, providing natural areas and wilderness. Threatened, endangered and rare species will be protected in order to survive and prosper.[1]

Does this vision represent current reality? Is it a realistic description of the future? Can it be embraced by the diverse people of this region and beyond? To answer these questions and meet the challenges of developing a process for achieving this vision, we must have knowledge about both the current conditions and the forces that operate here.

The vision that most people embrace is, I believe, one that says we need to maintain and enhance existing patterns of land ownership and traditional land use in the region. The Northern Forest is the focus of attention and worthy of recognition because of the past history of management by private landowners here for multiple resources including wood, wildlife, water, and recreation. The heated debate that has been generated by concern for this land is a compliment to the stewardship ethic of our parents and grandparents. The best way to encourage and continue what is, in my view, the most successful, sustainable, and politically acceptable future is to allow the people of the area to use their ingenuity to develop a diverse and ecologically sound resource-based economy of wood products, agriculture, and tourism, which will better their lives and continue to satisfy the world's needs.

To reach this vision of the Northern Forest we need to preserve a sustainable resource for a sustainable rural economy. The basic problem we face as humans is a growing human population. As our numbers grow in the future, considerable increases in the demands on forest land will occur. Population pressures not only lower the quality of life for humans but also contribute to the degradation of habitat for plant and wildlife species.

The regional or global aspects of population growth reach far beyond the scope of this discussion. The ramifications, however, are so integral to the challenges we face that we must keep our role in the balance of life on this planet in mind at all times. Since the 1700s we have reduced the world's forested land areas from about 60% to about 31%.[2] During this time the world's population has grown by 11 times.[3] In the United States, the forest-land-to-person ratio was 30 acres per person in 1750; today it is under 2 acres per person.[4] Our environment's limits can be stretched to some extent by science and technology, but the ultimate carrying capacity of our world is a limit we cannot escape.

The management of the forest for resource extraction and ecosystem integrity, as well as other noncommodity uses, is the critical issue for many of those interested in the Northern Forest. Regardless of whether you view forest management as having a positive or negative effect on the Northern Forest, the reality is that the forest is the predominant economic factor in the region. The economic and ecological viability of the Northern Forest will be determined largely by the future of the land as a resource base for the forest products industry.

As a resource base the Northern Forest is diverse, productive, and unique, but its future will be shaped by a world economy that functions far from the echoes of the trees that fall in the Green Mountains. In the last few decades the world has seen some of the most dramatic changes in history. As the world's sociopolitical structure continues to change, the global demand for forest products is expected to rise dramatically.[5] At the same time, the world's environmental organizations have successfully alerted the world to the major problem of global deforestation. Consumers have reacted to these ecological threats, and today we see major global reactions in forest products markets. Even though the contribution to global deforestation from timber harvesting is small (the bulk is attributed to slash-and-burn conversion for agriculture),[6] many nations have instituted import or export bans on tropical lumber and logs. The effect of this on the world's lumber markets has been more demand for northern hardwoods.

Here in North America the effects of dislocations in regional timber production have been exacerbated by regional changes. The western United States is undergoing an "old growth/spotted owl timber lock-up." Our National Forest system has been virtually removed from the timber production network by an organized series of timber-sale appeals from the preservationist community.

These global and regional changes create both positive and potentially negative effects on our predominately privately owned northeastern forests. More demand for our wood can bring more value to our presently undervalued forest resources. Also, this demand could lead to problems similar to those faced in other regions around the world. How can the citizens of the Northern Forest and the ecosystems in which they function adapt and prosper in the face of global change? Can the Northern Forest ecosystem provide for the demands of the world's population? What is the proper role for the Northern Forest in the global arena? The answers to these questions will set the course for the future.

I have briefly outlined some of the issues that affect the Northern Forest. We are part of a global economy. We also have an important global responsibility. The United States is the only economic superpower left in the world that still has abundant natural resources. We have a serious moral obligation to export to the world's developing nations both our technological knowledge and our stewardship ethic. It took our country many years of hard lessons to learn the drawbacks of resource exploitation. Yet we have been fortunate. Our forests have gone through decades of overcutting and poor management, yet have made incredible recoveries and are now among the most productive and well-managed forests on Earth.

The facts are clear. Despite the highly publicized reports that we are running out of trees and forests in the United States, the opposite is in fact true. Consider the following: over one-third of the United States is forest land. These forests contain 23% more timber now than in 1952, with over 6 million trees planted every day. There are 70% more hardwoods in this country now than 35 years ago, and softwoods have increased in all areas except the Northwest. The United States grows each year almost twice what we use, and there are over 94 million acres of wilderness on which timber can never be harvested.[7]

With modern advancements in technology, the forest products industry has become cleaner and more efficient in the value-added portion of production. New pulp and paper plants are geared toward recycling and reduced use of chemical treatments. Lumber processing, despite foreign competition, has improved both product utilization and energy efficiency.

The average person has little knowledge of just how successful and beneficial the forest products industry has been in the United States. Managing forests for the production of timber is not just a profit-motivated activity of the timber industry; it is globally responsible resource management, which produces an environmentally superior, renewable product. There are no substitutes for forest products that are more environmentally friendly. Plastic, steel, concrete, and aluminum all have high environmental costs. They are increasingly recyclable, but not renewable. I would much prefer to have lots of land that is growing trees to be used for wood products than to use other, more harmful, products.

The Northern Forest can play an important role by being an example to the rest of the world as a sustainable rural economy. In developing countries there is very little forestry planning or, for that matter, any forest-use infrastructure. By exporting the system of land management and forest products processing that we have here in the Northern Forest to other rural areas, we can improve the quality of life for other people and save other forest ecosystems from destruction.

Even grade school children now know that if people continue to harvest all trees from all areas and put the land to other uses global biodiversity will suffer greatly. Trees are cut down in many countries because they have no value. Managing forests to produce wood that has value is environmentally superior to other uses of the land.

The current trend of environmental paranoia over forest use is disturbing. Society is bombarded with controversy over what is the proper use of our "natural" areas. Many insist that we know too little of how ecosystems work to be harvesting trees. The fact is that we will probably never know everything. Instead, we must be careful and prudent and strive for knowledge by doing. It is not irresponsible to use resources in a manner that we know is beneficial to wildlife, people, water, air, and soil. The irresponsible action is for a nation as rich in resources as ours to use amounts of wood that we are not willing to produce or to reduce the supply of nonrenewable resources through the use of less ecologically compatible substitutes.[8] As a nation rich in resources and economic prosperity, how dare we dislocate the global demand for wood products to areas that are far less capable of prudent management, or to ecosystems without the ability to respond to resource extraction?

I envision the Northern Forest as a model forest, the best example of how a rural economy based on forest land resources can be successful and sustainable. I am an advocate of forest management and multiple use. Human beings are a critical component of Northern Forest ecosystems, not a parasite. We have a duty to control our use of and impact on the forest. Our role in this ecosystem has been shaped by history. Our role in the future is shaped by our actions today. Some involved in the debate over this region's future feel strongly that we should set ourselves apart from nature and that we should let the forests return to their "natural" state. Is this possible? What is "natural"? Is it responsible for us to leave alone what we have changed so much over the centuries? The choice is clear for me. It is not a matter of whether or not we use our forest resources but to what degree.

At our present level of socioeconomic development in the eastern United States it is unrealistic to think that we can successfully turn back the clock to a 1700s type of subsistence lifestyle. As romantic as it may seem, I believe that society is simply incapable of sacrificing that much. It therefore becomes even more imperative for the current trends of population expansion to be changed. We have too many people who use too much "stuff." Although I hope that some of the environmental consciousness that has been raised in recent times will result in advancements in recycling and product development, population growth and a consumption-oriented society mean that the Northern Forest is unavoidably going to see substantial demand for forest products and development pressure in the future.

In 1987 a United Nations commissioned report made popular the term "sustainable development."[9] This report reinforced the idea that economic growth is necessary for environmental protection: The needs of the present must be met without compromising the ability of future generations to meet their needs. The key to achieving this future will be the type of economic development and not necessarily the extent.

In my opinion, the forests in this region can be a sustainable resource for a sustainable rural economy. The answer to how we keep the Northern Forest sustainable must incorporate all of our knowledge of scientific, social, and economic factors into a broad-based approach to resource planning. Resource management decisions should be based on the capabilities of forest ecosystems to produce sustained, multiple benefits as well as on the social and economic needs of the region's communities. Understanding these two issues will help to narrow the many choices available in our quest for a desired future.

We should first address the social and cultural needs of the Northern Forest region. The environmental success of this ecosystem hinges on the activities of its human component. Rural economic development can enhance biological diversity in forest ecosystems. Multiple-use forests can meet the needs of the human inhabitants far better than a combination of preservation land and overdeveloped land.

A look at the part of the Northern Forest that I am most familiar with illustrates the case for my vision. Northern Vermont's portion is about 2 million acres in size and is comprised of our most northern counties. This area is very forested and very rural. Vermont is the most rural state in the country and, based on the 1990 census, this part of Vermont is the most rural part of the state. The forest types here are predominantly northern hardwood and spruce–fir. The economic base is a mixture of forest products manufacturing, agriculture, recreation, and tourism. Population size in most towns is small. Over 80% of the land is forested. Many of the families have lived in this area for generations. Almost all of the land is privately owned, and many of the landowners have carried the costs of stewardship for many years. Only about 15% of the area is owned by large forest industries, with even less held in state-owned forests and parks.

The forest products industry in the area is mostly small in scale. Many small to mid-sized sawmills make up the production base. The industry, however, is quite dependent on the vast infrastructure that exists throughout the region. Pulp mills, paper mills, fence post manufacturers, sawmills for a wide variety of tree species, veneer mills, log exporters, concentration yards, firewood processors, specialty manufacturers, and home-heating woodchoppers are all regionally important markets. The existence of this diverse infrastructure is needed for a value-added economy to function. On the surface it would seem that the existence of so many markets within and outside the area would detract from local value-added processing, but it does not. If we had just a few, rather than several, local markets for the wood that is produced locally, the lack of competition would keep prices low and even the smallest of logging operators would work themselves out of a job very quickly.

The Northern Forest in Vermont is diverse and productive. The Forest Service and State of Vermont Forest Survey show that we only cut about one-third of the annual growth.[10] Even though the harvesting rate is far below the growth rate, the value of the harvest is far from insignificant. No statistics are available for the 2 million acres in northern Vermont alone, but Vermont's statewide economic information offers some perspective: In 1990, Vermont's sawtimber harvest was 220 million board feet and 330,000 cords of wood, which brought in over $18 million dollars to landowners. Loggers and truckers received over $48 million in payroll, and secondary processors paid out over $128 million in payroll while adding $350 million in value to forest products. Total product shipments were valued at over $662 million.[11]

The economic importance of the forest is not small. These jobs are the ones most often held by native Vermonters. The culture of the Northern Forest is dependent on the land for its economic prosperity. The lifestyles of these people are far less consumptive and wasteful than their urban counterparts.

Yet the rural economy of the Northern Forest has much room for improvement. Many of the residents have incomes that are at or below the national poverty level. Work in the woods or in the recreation industry is seasonal. Service sector jobs are low paying. Clearly, diversification of economic opportunities is needed. Yet most residents are well satisfied with the lifestyle. The "voluntary poor" are the majority.

For years the value of unfinished forest products has been very low compared to their relative worth when finished. This must change. As with farm products, the American public has been spoiled by the success of productivity. The people of the United States have been subsidized by the rural landowners and workers for too long. This imbalance of payment will be reversed if our system of free enterprise is allowed to function properly. Many roadblocks still exist for a proper level of economic prosperity to reach the true stewards of the land. Society is going to have to pay for what it has taken for granted for too long.

We also function on an unlevel playing field when it comes to international trade. The close proximity of foreign markets for forest products is both good and bad. The infrastructure of Canadian and other international markets helps diversify product marketing, so its existence is critical. But there is a serious imbalance in the ability of local markets to compete with a region where energy, insurance, health care, and timber inventory are subsidized. Some overseas markets have been instrumental in raising the value of certain raw materials—and hence the value to landowners and workers—but they continue to draw substantial amounts of value-added potential out of the area. The low cost of labor and lack of other overhead in these countries enable them to compete effectively—yet unfairly.

Although many problems still remain, the potential for great economic benefit exists for the communities of the Northern Forest. The resource is available and can be used sustainably. The advancements that are needed in rural development and economic diversification can be accomplished.

The people of the region also need to be empowered to reach their potential destiny. By and large they are strong-willed and independent. Because of a historic lack of practice, most of the people here tend to lack the consensus-building skills and political experience that is needed to control the outside forces that now threaten their traditional lifestyles. Their potential prosperity is now threatened by the high-profile national interest that has come with the Northern Forest Lands Study. There are only about 700,000 people that live in the entire 26 million acre Northern Forest land area of the four states. Yet within a day's drive there are over 70 million people.[12] The political power of the native inhabitants to control their own destiny is severely limited by both numbers and cultural experience.

It is very easy for someone who lives in Boston or New York to sit in a condominium reading *Audubon Magazine* and develop a highly romanticized view of what life is or should be here in the North Country. The simple act for that person of writing out a fat check to one of the powerful preservation organizations or zipping off a letter on a personal computer to a congressperson is unfortunately the real political power in today's world.

The process of involvement in the Northern Forest requires that everyone's viewpoint be considered. It is hard when you have been in the woods or barn all day to drive an hour to some esoteric-sounding "public hearing." My point here is that the true stewards of the land, those who have safeguarded its ecological integrity, paid the taxes, worked the soil, and supplied the world with the products from the land, will now have to entrust their future to a public that understands little about the facts or the issues. The right of self-determination has been earned by decades of sound stewardship. Will the process recognize the value of local insight or will the propaganda wars between preservation and property rights obscure the view of the forest and its inhabitants?

What is best for the forest-dependent communities of the Northern Forest should be what those communities feel is best for themselves. The traditional lifestyles and ownership patterns have served the area well. There is a need for improvement, but a new level of economic prosperity can be reached without sacrificing the environment or the local culture.

One focus of much of the debate on the Northern Forest is over its health as an ecosystem. Can the forest provide for all of the commodity and noncommodity needs of society and remain a viable ecosystem? I believe the answer is yes, and there is a large body of evidence to support my contention. The eastern forests of the United States have gone through massive changes. Early European settlement resulted in extensive land clearing. The once "virgin" forests of New England are now predominantly second growth. Vermont has gone from being at the heart of the most heavily forested continent in the world before settlement to about 80% cleared land in the mid-nineteenth century. Presently it is almost 80% forested. This recovery shows both the resiliency of the forest and the extent of human influence through history.

The forest ecosystems of northern New England have been intensively studied for a long time. The history of scientific research in this area is one of the richest on earth. Many of the world's foremost educational and research universities are located close to this forest. The U.S. Forest Service has many of its highly respected Forest Experiment stations located in our eastern forests. Private industry has large landholdings in this area that have seen years of carefully conducted treatments for a variety of forest management directions. Cooperative efforts between private industry and research interests have generated huge volumes of information that is available to forest managers.

The biological diversity of this forest ecosystem has fluctuated along with the intensity of land use. The dynamics of the variety and variability of forest organisms and the ecological complexes in which they occur will always be dependent on the level of socioeconomic development of the human component.

The bulk of modern wildlife habitat research and biological study indicates that a mosaic of forest and nonforest habitats best meets the needs of the widest variety of species. The current pattern of land use that is present in northern Vermont has many advantages for a wide variety of wildlife. The experts in the field of biological diversity support the maintenance of diverse forest types. Forest management is the only way to maintain the successful abundance of habitat variety that exists in the Northern Forest today.[13]

The issue of what we do or do not know about forest ecosystems continues to be brought up in the Northern Forest debate. In Vermont, at least, we have made major steps at looking very carefully at how the human resource extraction component of the Northern Forest ecosystem effects other components. One of the most recent studies was done by the University of Vermont on the effects of timber harvesting.[14] This was an in-depth inventory of the impacts of timber harvesting on a wide variety of values. The study looked into both the present characteristics of timber harvesting and the impacts on aesthetic values, archaeological resources, threatened and endangered species, timber quality and productivity, water quality, and wildlife habitat. Some of the highlights include (a) involvement in some way by a professional forester or wildlife biologist in 77% of timber harvesting operations, (b) harvesting contracts in over 70% of the operations, 82% of these with water quality or soil erosion provisions and 77% with provisions for aesthetics, (c) silviculture as the stated objective on more than 83% of operations, (d) selection cutting on 81% of operations, and (e) clearcuts in only 9% of the operations. The negative effects on the variety of resources were minimal; in many cases the effects were strongly positive.

Based on field examinations, specific practices to enhance wildlife habitat were evident on more than half of the operations. No detrimental, short-term impacts on wildlife were identified in the 78 timber-harvesting operations studied. Rather, a high proportion of operations exhibited conspicuous management practices that indicated a desire to improve wildlife habitat. Yet the study results have been criticized by some and ignored by others in the antiharvesting "club" because it did not show them what they wanted to see. The recommendations of the study group contained the following statement:

> [I]t is the unanimous recommendation of the study team that new legislative initiatives designed to regulate timber harvesting in Vermont are not justified at this time.[15]

This study should have helped to improve the government's and public's level of awareness and respect for the role of timber harvesting in Vermont. The fact is that so far this important research has only succeeded in derailing poorly drafted, unnecessary legislation.

The importance of the forest products industry in the sustainability of the Northern Forest's economic and ecological future is still undervalued by those who make public policy. The critical role played by the Northern Forest in preserving large-scale regional biodiversity is not understood by today's lawmakers. The continuing debate over the proper use of timber harvesting will be kept an issue by the small but vocal groups that are philosophically opposed to any consumptive use of forests at all. Some debate is appropriate and useful. But when the potential for true progress in sustainable rural development is damaged, all citizens of the earth suffer. Responsible stewardship of the land is required of all persons, not just landowners.

My view of the Northern Forest as a sustainable resource is based on the fact that the resource is presently being managed in a sustainable manner and that we can maintain a high level of ecological integrity at even accelerated levels of use. The rural economy is also sustainable with ample room for expansion at the current level of socioeconomic development.

The process of sustainable development in this area faces many challenges. Preservation of the rural character and cultural heritage will be a major challenge. The forces of change that will affect the area in the coming decades can be managed. What will affect the cultural and biophysical landscape will also shape the ecosystem and our ability to produce livelihoods without harming that ecosystem.

The interconnectedness of the four-state Northern Forest region is critical for factors such as forest-utilization infrastructure and regional biodiversity. The uniqueness of conditions in each state is, however, demonstrated by their particular land-use patterns and main political issues.

Northern Vermont is owned predominantly by small, private landowners, and the main issues involve taxation of open land, property rights, and the role of land-use regulation. Northern New York is dominated by the Adirondack Park, which includes large areas of public wilderness and controls on the use of private lands. The major issues there are the limits on land use and development, future public acquisition, and a very strong grass-roots backlash against centralized, top-down planning, which has resulted in protests and isolated violence. In New Hampshire, the small area within the Northern Forest boundary is a mixture of public land within the White Mountain National Forest and some large landholdings of the larger forest products companies, and the major issues there are public land management (many appeals of public timber sales have angered many in the wood-using community) and the future of large corporate ownerships. The North Maine Woods is a vast area of mostly large, corporate land holdings. The forest products industry is the major economic force in the area. The issues include large-scale clearcuts by forest industry, ownership of very large parcels of land by multinational corporations, interest by national preservation groups in large-scale acquisition of land for wilderness, and continued access for public recreation.

It is important here to note that the land ownership pattern and land-use practices of Maine are critical to the question of why the future of the Northern Forest is still an issue. I believe that large industries have contributed greatly to the quality of life in northern Maine. The area's system of recreational access to private land should be seen as a great value. The private landowners who provide these benefits to the public should be rewarded as stewards, not punished by the "taking" of these rights through some regulatory process. What would the landscape of northern Maine look like if it were not for these large landowners? Would there still be the same amount of undeveloped land or lakeshores? What would the economy of the area be like? Is the use of this land to produce vast amounts of wood fiber for the world's demand the highest and best use for our global community?

My own opinion on this issue is that the pattern of land ownership in the North Maine Woods has provided years of valuable benefits, from forest products to recreation. The global pressure on the long-term economic stability for the large corporations is very disturbing. I feel it could be very damaging to the resource for wood production, wildlife, and recreation if major changes occur. These changes would be detrimental no matter which way they go—either the liquidation of timber and landholdings for short-term profit or the sale of large parcels to the public for preservation. Both scenarios will change the values now available. What is needed is long-term stability. Society must strive to create an economic climate that supports the long-term ownership of forest lands by a variety of private owners for multiple-use resource conservation.

The debate on the future of the Northern Forest has resulted in the discussion of many different options on what the best strategies are to protect what is valued. There is not much debate over the values to be protected. There is enough variety of opinion over what people like about the area that all of the values are important. We will continue to argue over whose use should come before another's, how much of the resource should be allocated to what use, and who should pay. The simple answer is that there is no simple answer. It is going to take a very broad, complicated process of planning, economic incentives, tax structure changes, and infrastructure support to reach a sustainable future.

The many ideas and strategies that are being forwarded all have potential to contribute to the success of the Northern Forest. I believe that a major problem that we face is that those who feel very strongly on one extreme or another will detract from the needed combination of strategies that must be implemented.

The emphasis that is given to any one of the strategies being pushed by each special interest group will be critical to the future we will shape. The advantages and disadvantages of the different politically driven agendas are critical to my view that something different altogether is the answer. I advocate a broad-based approach to maintain more of the Northern Forest, a balanced combination of many types of conservation programs and initiatives. We must set priorities and tactics that will effectively produce the most economically and ecologically sustainable future, and we need to do this now. The process, however, needs to progress slowly, in a manner that creates a manageable environment for existing industry and culture.

My intent in the remainder of this chapter is to outline many of the components that will play a role in the integrated approach to resource conservation that I support and to discuss what I feel the extent of each component should be, as well as their strengths and weaknesses.

Ownership of the Resource

One of the highest priorities should be to maintain and strengthen existing ownership patterns. Public ownership has a role in the protection of resources. But the limited amount of public land that is now present in the Northern Forest continues to demonstrate the problems associated with public ownership. The management costs are many times that of private land. The pressure from special interest groups to manage all of the land for single-use purposes is increasing. The tax loss to small communities can be burdensome. Often the mandate to try to be everything to everyone results in inappropriate management. Wilderness is designated by a political rather than a scientific process. The "enhancement factor" that occurs on any land that is adjacent to public lands is a major force for land fragmentation. The attractiveness of these adjacent lands for recreation, investment, and second home development is very high, which drives up the price of land for forestry use by private landowners and industry. The National Forest acquisition process itself is a problem. The Forest Service will pay prices for land based on appraisals that are driven by sales of land in the areas adjacent to the National Forest. These prices are far higher than what is a reasonable price to pay for land if you need to produce an income from that land. Finally, the worst factor is that we abdicate the management of the land from the professionals to the politicians.

Despite my criticism of large public ownership, I feel that it can play a limited role in resource conservation. Certain unique and fragile areas may be best protected by public acquisition. When this is appropriate the most local level of public ownership is best: local land trusts or town/village forests, then conservation organizations or the state, and, as a last resort, the federal government. However, until the National Forest system can resolve the problems it has with achieving reasonable multiple-use management that is affordable and not subject to micromanagement by special interest groups, it is not a welcome player on the Northern Forest conservation team.

The role for conservation of forest lands by public entities should also involve the smallest amount of rights necessary to achieve the protection of the resource. The use of conservation easements and purchase of development rights is far more palatable than outright purchase. The use of acquisition of easements and other less-than-fee purchases has potential. This option needs to be tried to see if it is viable for long-term protection that is not too intrusive or limiting to the landowner. The taxation question still needs to be resolved in this area for the ownership of land for long-term forest use and productivity to be feasible.

The idea of any large-scale National Forest or National Park for the Northern Forest area should be dismissed outright. This would be an irresponsible use the public's money—even if it were ever available! The effects and ramifications of this scenario are very disturbing. The decrease in the diversity of land-use patterns would be biologically devastating. The costs both economically and culturally are staggering. The effect on the economy of the area would affect the global economy. The sociological changes from an independent, production-oriented work force to a dependent, service society would indeed be cultural genocide.

The best way to conserve the forest is to keep it in the hands of those who use it, understand it, and love it.

Land-Use Regulation

The level of land-use regulation varies widely in the Northern Forest. It appears to me that the areas with the most limited land-use controls are the areas with the best record of stewardship. Strict land-use controls tend to stifle economic progress. A great divergence in economic prosperity can arise between those who work for the regulating body and the citizens that depend on the land for their incomes.

Centralized planning may be considered by many as successful, but for the rural landowner and worker it is usually devastating. We need only look at the Adirondack Park to see what can occur. Nowhere else in the

Northern Forest has so much hostility developed between the governmental entities who administer land-use controls and the residents of the area. People see their property rights more and more limited, and they resent the continued use of public funds to buy land that will be forever removed from the production of commodities.

The issue of private property rights versus public interest is a major problem that must be resolved for the future of the entire Northern Forest. At issue, regardless of the level of land-use control regulation, is what rights a landowner has and at what point is the public owed a level of protection. Many court cases will have to be tried before this issue is totally resolved.

There is definitely a lot of room for debate on this issue. The Constitution clearly gives rights to private property owners, yet those rights are not without limitations when it comes to the "public good." Clearly, landowners must exercise their rights with the understanding that the public also has rights to health and safety. How does this relate to forest land? Do any rights belong to the land itself? Are there rights for creatures of the earth other than humans?

My feelings are that the land has inherent worth and ecological values, and that it has rights. We have a duty as landowners to exercise our rights in a manner that safeguards the rights of the land and its integrity. The public also has a right to a certain level of protection of ecological values on private land. In regard to the Northern Forest, this is a critical factor. The land and its values should be protected by the legal title holders for future generations. To what extent? At what cost? And who is responsible for carrying that cost? Is it the landowner or is it the public and society as a whole that should be responsible for the costs to landowners for maintaining ecological values?

There seems to be a tremendous double standard when this issue is put in the context of forest land conservation and ownership. The landowners who have safeguarded the land's undeveloped condition are now put at a disadvantage and treated differently than landowners of developed property. Do lands that are now developed not have "rights"? If society is to require forest landowners to safeguard things like public access, wildlife habitat, water quality, and scenic views, then why not suburban or urban land owners? How would the condominium owners of Burlington or Boston like to let me "camp out" in their living room when I go to town? Should suburban homeowners that have a spare bedroom be required to let the homeless stay over? Should we ask owners of expensive homes to tear them down and replace the wildlife habitat that once existed there? How will society interpret the Constitution with regard to what rights are affected and who gets the "grandfather clause" and when?

These examples may seem extreme. The fact is that if we are to protect and enhance what is left of our "natural heritage" it is vital that it is done in a manner that is fair and equitable to the owner who has an investment in the land. A system of incentives and rewards for good stewardship is the answer, not restrictions and penalties. Again, the costs of protection should be borne by all of society, not just the rural landowner who happens to hold the property at this point in history.

The history of private ownership is so deep in New England that I believe major initiatives for land-use control will not be acceptable or politically achievable. Nor do I think they are appropriate.

Taxation

The system of open-land taxation that we now have in most of the Northern Forest is regressive and detrimental to long-term conservation. In Vermont, land is taxed at its "fair market value," which means that open land—forest or agricultural—is appraised based on comparable sales of similar properties. Properties are usually sold for uses other than their present "use value." When land is appraised at values for development or speculation, it is valued far above its value for producing forest or farm crops. The result is that the tax burden on the land exceeds its ability to produce income.

Throughout the Northern Forest region, land is bought and sold for many reasons. This is good for economic productivity, and the sale of land is itself not a problem. Forest products companies buy and sell land as part of their business. We cannot and should not control the sale of land. What we must do is give an advantage to investors who wish to hold land for long-term forest use. The present system of land taxation does just the opposite.

Many states presently have programs intended to ease the property-tax burden on farm and forest owners and make the system more equitable. The problem is that these programs are subject to swings in public funding and political tinkering. The key is to have a system that is consistent and reliable. Long-term planning for investment in forest land requires stability. The issue of taxation may currently be the single most important key to resource conservation.

Economic Infrastructure

The economic productivity of forest land will determine if the land remains in an undeveloped state or is put to other use. Society must realize the multiple benefits that it receives from forest land. Land that is managed for timber on a long-term basis produces a wide variety of benefits for the public. The clean air, abundant water, wildlife habitat, scenic views, and recreational opportunities that the public enjoys rarely bring any economic benefit to the forest landowner. If society is unwilling to factor in to the economic equation these attributes, many undesirable things could happen. Some landowners will be forced to convert the land to other use. Others may sell their investment, sometimes after liquidating the timber assets.

Large landowners are finding that they can charge for the nontimber values they produce. Hunting and other recreational leases are becoming more common, usually as exclusive leases for wealthy groups. This is a reasonable option for a landowner who is given no incentives to provide hunting or other recreation opportunities to the public. This trend toward fee recreation is understandable yet disturbing. Do we want a system that requires wealth to enjoy the forest? Will the tradition of public enjoyment of open land be destroyed by a society that is so short-sighted it fails to give incentives to forest landowners who provide public benefits? The exact type of incentives is not as important as their effectiveness. Different ideas may need to be tried before the correct "fix" is arrived at for the long term.

What will keep the resource sustainable is a stable, industrial infrastructure. Forestry is a long-term business. The economic climate must be made more predictable. The factors that need to be stabilized are many, and include tax policy and the regulatory environment. I do not mean that there should be fewer regulations, but that for long-term planning, industries and small businesses alike need to be able to predict options and costs.

Ecosystem Health/Biodiversity

Society must be made aware that we can have a healthy environment with continued resource extraction. The economic stability of a forest-based culture depends on a healthy forest. Functioning forest ecosystems will only occur when there are abundant amounts of forest land. To ensure that the most possible acreage is kept in a forested condition, we must make the ownership and management of forests profitable for private owners.

The forest ecosystem is not something apart from human life; our culture is a part of the forest. It may be labeled by some to be an anthropocentric viewpoint, but people are the most influential component of the global ecosystem. It is our duty to control our impact as best we can given our skills and knowledge.

The issues of wilderness and remote wild land is one that must be addressed for the vision of the forest to be achieved. There is a great deal of public support for wilderness preservation. Does the public truly understand what it wants? Many do not know what true wilderness is. Many remote places are considered "wilderness" by those who visit, yet they have been managed for timber production. Is more wilderness appropriate for the Northern Forest? If this means large areas of land restricted from harvesting, I would say no! There are still some special areas in need of protection from human use, but we cannot afford to limit our future too greatly. The same values of remoteness and solitude can be incorporated on multiple-use lands. Less intensive harvesting and longer rotations can achieve the same ecological and social values that true wilderness can provide.

The forest ecosystems of the Northern Forest require a high level of involvement by humans to remain healthy. The impacts of global travel and trade have brought a variety of introduced elements to the environment that now must be managed. It would be irresponsible for us to now turn our backs on nature. Just think of the issues of ozone depletion, El Niño, global warming, and acid rain. The effects of these forces are not yet completely understood. It is not sensible to limit our management options too severely.

The forest insects and diseases that have been introduced by human activity in the last century continue to influence the forests of this country tremendously. Many contend that this is "natural" and should be allowed to occur "naturally." They are "natural" in my definition because humans are a part of nature. So it is also "natural" for humans to control what we have caused. Gypsy moth, beech scale, pear thrips, hemlock wooly adelgid, balsam wooly aphid, chestnut blight, Dutch elm disease—all of these have had, and continue to have, dramatic effects on the forest. All of these were introduced by human activity.[16] Would it be "natural" for us to allow our forests to face these threats without our help?

Wildness and the benefits of large "undisturbed forests" have a place in some parts of the world and other parts of this country. Here in the Northern Forest the role of wilderness must remain very limited. The allocation process must not be political.

Planning and Research

The complicated, integrated approach to forest conservation will need extensive research and planning. No vision for the future can be achieved without the tools and a plan on how to use them. Most critical will be the participation in the planning process of the people that control the resources: landowners, woodsworkers, politicians, activists, and John & Jane Q. Public.

This planning process must be accessible and user-friendly. We have a long way to go to make this happen. The infrastructure is in place, but the average citizen is either apathetic or has been turned against planning by mistrust and past abuses. Local and regional planning for forest resources and economic development must become socially acceptable. This will occur if planning is allowed to be truly bottom-up without the baggage of state and regional authority. Centralization of the planning system will create public resentment. It also will promulgate an ineffective and insensitive bureaucracy. Landowners who feel that they are being told what to do with their land will fight planning all the way. If they are made part of the process they will feel in control of their destiny.

Even the identification and inventory of resources, an element that should be critical to sustainability, cannot be achieved without the assistance of local landowners. When the local people feel that they truly have self-determination, the process of long-range resource allocation and planning can proceed.

Without a process to face the threats to a sustainable future, the citizens of the Northern Forest are in danger. The threats are real. The urban/suburban interface is advancing northward. People want what we have: open space, a rural quality of life, and country living. The success of the tourism industry helps to diversify our economic base. However, with it come people who want a piece of the land for themselves. Fragmentation, higher land values, and more people are the result. With the people come the urban attitudes that are far less tolerant of consumptive forest use. Logging and hunting seem messy and bad for the land. Rural values must be respected and maintained as crucial dynamics of this forest ecosystem.

The need for local planning is real. It can mean better economic opportunities. Property rights can be protected and even enhanced. The biggest benefit is the power of self-determination and some influence on the future of our culture. We must continue the allocation of financial resources to education and scientific research. New studies on ecosystem health and forest utilization should continue and expand. The integration of many fields of study must be applied to forest management. Advancements in how we manage multiple values are coming all the time. The Northern Forest has great resources for research and education. The importance of forests and their role in life on this planet will drive further advancements in technology and environmental ethics.

An important role for the National Forests and some other public lands should be research and the demonstration of the best management practices available. New ways of blending ecological protection and forest productivity need to be demonstrated. Public lands are well suited for this work. The costs can be borne by the public. This is fair and appropriate so that the education of private stewards will be accomplished. Society must contribute to the benefits we receive from private forests.

Resource Identification

Planning for the future will require knowledge of what we have and how long it can last. The gathering of information is necessary to plan for the needs of our society, and so that the stewards of those resources can be rewarded for their efforts and start to receive a proper return on the values they provide for society. The categories for resource identification, as outlined by the New England Society of American Foresters, include wildlife habitat, scenic and recreation areas, riparian and aquatic zones, productive timberland, areas with unique communities of threatened or endangered plants, lakeshores, contiguous blocks of forest land, historic areas, and wetlands.[17]

To both utilize and conserve our natural resources will require information on the availability, amount, and capability of these resources. Society will demand more and more from our forest resources. How we use what we have will determine what future awaits our culture and the global ecosystem.

Education and Political Action

Of all the factors and strategies that we must consider, none will have as profound an effect as the political process. The coordination of a broad-based, integrated program for forest conservation will be difficult. To coordinate the many strategies involved will take a knowledge of both the problems and the proper solutions. It will also take the desire to make the solutions happen.

To make the proper process occur the political system must set in place the changes that are needed. Policymakers and the voting public must understand what is best for the Northern Forest. The process of public education must progress at a faster pace. Too many people still do not understand just how limited our choices really are. Disagreements tend to be driven by opinion rather than by knowledge.

Fewer and fewer people have a true link to the land. When people are no longer dependent on the land for daily survival, they tend to take for granted what the land gives to us. Connection to the forest brings understanding of our role in the forest's future.

Summary

The Northern Forest is a complicated ecosystem undergoing change at the hands of its most complex component: human culture.

Our present society has the resources to develop a sustainable culture with respect to the forest. We have sufficient data to show us that the present resources of the forest can meet the present demands of society. What of future demands? How will society change? How will the forest change? Can most of this ecosystem's inhabitants survive together long into the future?

The challenges will continue. The Northern Forest will have to adapt. The most adaptable component of this equation should be humanity. Can we? Will we? We can continue to learn about our role in the forest, to use the forest wisely, and to manage our impacts to the best of our ability.

The action plan for building a sustainable future must have a starting point and should have a list of priorities. Debate and dialogue are fine, but the time for consensus-building and action is now. Humanity will always demand more of everything. Even so, the level of sustainability that we need to reach may be more achievable than we think. If society could take a little more time appreciating what we have and where it came from, we would understand our link to the forest far more clearly.

The forest and the environment have become very popular in recent years. There are all manner of tree lovers and forest activists. Activism will bring awareness of the forest to the public, but not necessarily knowledge. What is best for the forest must still be driven by the integration of compassion and scientific reality.

11

Northern Appalachian Wilderness: The Key to Sustainable Natural and Human Communities in the Northern Forests

Jamie Sayen

We abuse land because we regard it as a commodity belonging to us. When we see land as a community to which we belong, we may begin to use it with love and respect.—Aldo Leopold, *Sand County Almanac*, 1949

Most of the debate over the uncertain future of this region currently focuses on propping up the status quo. While this may appear to be politically expedient, it is premised on the uncritical acceptance of the prevailing assumptions that the status quo has served the region well and that environmental and social concerns must not hinder society's "need" for growth, development, and resource extraction.

The following essay is premised on an entirely different set of assumptions: (1) that a sustainable economic, social, and political system cannot be based upon ecological abuse, (2) that critical ecological, economic, and social issues must be addressed on an appropriate scale—a local or regional level whenever possible, (3) that ecological and economic recovery are inextricably bound together, and (4) citizens of this region share a deep and abiding love for the region that far transcends the differences (such as jobs versus environment) that have traditionally been exploited to divide the community against itself.

Vision

To live sustainably, we need to recognize that protecting the forest means protecting the human communities that the forest sustains. Sound economics means protecting your capital. Our "capital" is the soil, air, water, forests, wildlife, and human residents of these Northern Forests.

How can we assure sustainable natural and human communities? This can be done by focusing our efforts on four key issues:

1. The habitat needs of all species native to this region (including species that have been locally extirpated) must be assured.

2. Human populations and consumption patterns must not exceed the ability of the region to meet our needs and desires. Human settlement must complement, not conflict with, the habitat needs of nonhuman species.

3. Human economies must be ecologically benign and sustainable, provide for basic human needs in a socially responsible manner, and rely on renewable resources that are managed on a sustainable basis.

4. There must be social equity within generations and between current and future generations.

Crisis

The Northern Forest region is not homogeneous. Each state is quite different, and there are different regions within each of the four states. Each community in the region has its own concerns. These differences must never be overlooked, but they must not blind us to the realization that the communities of the four states have much in common and face many of the same threats.

The Adirondack Park, where 42% of the park is publicly owned as "forever wild" state forests, is threatened by land speculation, subdivision, and second home development on private lands. Development also threatens shorelines and other critical tracts of land in northern New England. The gravest threat to the health of the privately owned forests of northern New England, however, comes from unsustainable forestry practiced by largely absentee corporations and large timber contractors.

Today, the forests of the Northern Appalachians are, by and large, in the worst condition they have been in since the retreat of the ice more than 10 millennia ago. There is virtually no old-growth forest. In addition to the many extirpated and extinct native species, many subspecies, species, communities, and ecosystems are declining dramatically. Neotropical migratory songbird populations are decreasing. Amphibians, some of which are acutely sensitive to clearcuts, are threatened globally. Erosion, desiccation, pesticides, and nutrient loss are destroying the source of terrestrial life—the soils. Our rivers are dammed and polluted. Global threats to the region include the greenhouse effect, the growing hole in the ozone shield, and acid deposition.

And now our region's economy, dominated by transnational paper companies, is in crisis. All three of the Coös County, New Hampshire, paper mills were offered for sale in August 1990. Millions of acres of forest land owned by paper companies are for sale. In October 1991, 10% of Maine was sold in a single transaction.

Our human communities are in a shambles. From late 1991 to early 1993, over 100 jobs have been lost in the towns of Groveton and Stratford, New Hampshire; more than 300 have been lost in Coös County, which has a population of under 30,000. My neighbors in Groveton live in daily fear that their jobs will be terminated, that the mills may even close down. Poverty, drugs, alcoholism, and serious health problems afflict these towns. But an even more sinister disease is destroying our communities: a loss of hope. We are a divided and powerless community. Our destiny is in the hands of absentee owners.

The Legacy of Private Property Rights

There are two very different categories of private property owners: those who live in the region and have roots in the community, and the absentee, industrially oriented owners. These two groups often have conflicting interests.

When the Pilgrims arrived in 1620, they were appalled by the "hideous and desolate wilderness, full of wild beasts and wild men."[1] Later, Cotton Mather dismissed claims that the land belonged to the native peoples because "all knew [it] had been given the whites by the English crown."[2]

British business interests attempted to stifle colonial American manufacturing, warning Parliament "that the colonies' progress in iron and steel manufactures will promote the favorite scheme of America: independence."[3] The British wanted to retain the colonies as a source of raw materials and as consumers of products manufactured in England. Ownership patterns today are similar to those of 200 years ago, except control of this region has shifted from London to Philadelphia, Richmond, Boise, Atlanta, Purchase, Stamford, New Brunswick, and, perhaps, Tokyo.

In the post-Revolutionary period, speculators bought up vast tracts of the northern forests for a pittance. William Bingham bought a million acres in the Penobscot and Kennebec River watersheds in 1792. He led the fight against Maine statehood for decades. After the Civil War, Maine disposed of 3.2 million acres for as little as 12 to 30 cents per acre. The government disposed of its timberlands to pay its debts and to stimulate "growth."[4]

The story in New Hampshire was similar. W. R. Brown, long-time head of Brown Company in Berlin has written, "by 1867 the state had parted with all its timberlands at a small percentage of their subsequent worth."[5] The 1893 New Hampshire Forestry Commission Report deplored the fact that all New Hampshire's mountain forests "are private property and we have no more control over the curious treatment of them than we have over the moons of Mars."[6]

When the price rose, small owners sold, and auctions in the 1870s concentrated land in a few bidders' hands. By 1930, 32 people or corporations owned 85% of Maine's timberland.[7]

In the United States today, two-thirds of all privately owned land is held by 5% of the population.[8] Over 60% of the private land in the Northern Forest Lands Study area is currently owned by industrial or large non-industrial owners. According to the Northern Forest Lands Study, there were 14.2 million acres of private land in Maine's portion of the study region. The 18 largest landowners—less than 0.1% of Maine landowners—own over 75% of the Maine Woods. And yet, as W. R. Brown observed 40 years ago, there is "a general dislike in New England of absentee regimentation."[9]

The effect on communities controlled by absentee owners is devastating. Mutual aid and community self-sufficiency are replaced by a dependency on faraway forces. People feel powerless; passivity, apathy, and an abdication of personal responsibility replace self-esteem, pride, and hope.

The transnational corporations that own so much of the Northern Forest are often beyond the reach of local, state, and even federal governments. They are unaccountable to the democratic process. When organized labor or environmental groups struggle to enact legislation that protects the environment or workers' rights, the transnationals simply relocate their operations to a part of the country or world with less stringent controls, cheaper labor, and a more abundant—and less degraded—supply of raw materials. Free trade agreements, such as NAFTA and GATT, promote this corporate evasion of environmental and social responsibilities. As Helena Norberg-Hodge writes, "This is the meaning of the free market for transnationals—freedom from constraint in their search for new profits."[10]

The costs of absentee ownership are high:

The myth of Paul Bunyan. The logger and river driver of the nineteenth century have become mythological figures like the cowboy of the American West. But the truth was far less romantic. They were by and large landless, poor, hard-drinking men who were ruthlessly exploited by speculators, mill owners, and contractors, many of whom made fortunes. Smith writes,

"The work was hard, and there was little else to do. The crews, especially in the later period, were rather poor specimens, many of them. . . . The jovial lumberman can, and should be, confined to novels."[11] Smallpox and fatal fires were frequent visitors to logging camps. When a river driver fell into the river, his cynical mates would shout "Never mind the man, but be careful of the peaveys—they cost three dollars."[12]

High death and injury rates to loggers. Most loggers are paid by the piece rate, a practice that was abolished in the coal mines of the Appalachians in the 1920s. The accident rate in Maine logging operations is three times that for all other forms of manufacturing. In Sweden, where loggers are paid an hourly rate, the accident rate is on a par with other manufacturing jobs.[13]

Job loss. In both mills and in the woods, job loss due to mechanization and automation will intensify with the passage of time. In Maine there was a 30% reduction in logging jobs between 1984 and 1989. During the same period, 1,400 mill jobs were lost.[14] Nevertheless, the amount of wood cut nearly doubled in the past 20 years.[15] Between 30% and 50% of the loggers in northern Maine, New Hampshire, and Vermont are Canadian citizens.[16] Real wages for loggers have dropped significantly since 1970.[17]

Mill pollution. Pollution from saw mills has poisoned the region's rivers with dioxins, other organochlorines, and other toxic substances. In 1989 James River's Berlin mill emitted as much toxic air pollutants as the entire state of Vermont. The Androscoggin River is so polluted by James River, Boise-Cascade, and International Paper mills that Maine has pumped liquid oxygen into the river to restore dissolved oxygen levels so that fish and other aquatic organisms can survive. High rates of rare forms of cancer are found in mill towns. Although every major paper company in Maine has paid large fines for violations of pollution or health and safety regulations in the past couple of years, these billion dollar corporations merely pass the price along to the consumer as part of the price of doing business.[18]

Price fixing. The mills have forced smaller, nonindustrial owners to overcut their lands because they are receiving essentially the same price for a cord of wood that they received a decade ago. During this period their overhead costs (e.g., machinery, workers' compensation) have skyrocketed. Price fixing is a tradition in this region. Smith's history of the Maine lumber industry reports that there were accusations of mill price fixing and wage fixing as early as 1873–1874.[19]

Raw log exports. The export of raw, uncut timber has also been a problem for the region for over a century. In 1887 the *Bangor Daily Commercial* called for a ban on the export of round logs, hoping thereby to stimulate sawmills in the state.[20] Today there are 43 Canadian saw mills along the Maine border. About 80% of their saw logs come from Maine. Champion International admits that 85% of the softwood sawlogs cut on its Vermont and New Hampshire lands are exported raw to Canada, and 75 to 80% of these return to the United States as dimension lumber. A 1985 study estimated that raw log exports may conservatively be costing the Northern Forest region 14,000 jobs, $109 million of forest industry payroll, and $196 million of value-added revenue (in 1983 dollars).[21]

Sustainable forestry. Industry tells us its practices are "sustainable." They want you to think that this means the health of the forest ecosystem is sustained. What they really mean is that they can sustain the flow of wood fiber to their mills.

Actually, the timber industry has never practiced sustainable forestry on anything approaching a region-wide basis. If it had, we would still have old growth in the northern Appalachians. But in the early days, the attitude held that these forests' timber would last "until hell froze one foot thick."[22] Smith reports that an average sawlog in Maine contained 343 board feet in 1833. By 1857 the average was down to about 200 board feet, and by 1915, an average log contained a mere 58 board feet.[23] Shortfalls are predicted for most commercially valuable tree species in Maine by the end of the century.

Forest ecosystems have been degraded. Fragmentation and degradation of habitat by such practices as clearcutting, pesticide spraying, and road building are causing irreversible loss of biological diversity. Industrial forestry has caused soil erosion, soil degradation, nutrient loss, disrupted hydrologic processes, reduced wildlife habitat, and a decline in population levels of aquatic species. Insect epidemics, such as spruce budworm outbreaks, have been aggravated by such practices.[24]

Because industry has owned so much of the land, it has in some states traditionally controlled state government. There have been numerous efforts to regulate the timber industry, create public forests, and protect water resources since the end of the nineteenth century. Industry has scuttled or gutted most of these measures. Its control of the Maine legislature and governor remains unshaken.[25]

Claims that industry needs a "stable regulatory environment" ring hollow when viewed in the context of industry opposition to almost any regulatory effort. So long as industry prevents passage of regulations that protect the environment and human health and interests, pressures for these protections will assure an "uncertain" regulatory environment.

In addition to thwarting meaningful regulation of its practices, industry has exploited its control of government to secure numerous subsidies from the taxpayers. A partial list of subsidies includes cheap land purchases, fire suppression, spruce budworm spraying, investment tax credits, extremely low property taxes, government-sponsored research, an absence of effective regulation of forest management practices, and pollution privileges.

These subsidies do not benefit the taxpayers of the region. New jobs are not created. Benefits and profits leave the state. They do not promote real, sustainable growth. Real growth is an increase in the capital base, but the current cutting and polluting practices represent an accelerated depreciation of resources such as wood fiber, soil, water, fish, and petroleum. This is a debt-based economy, and the debt is growing.

Opportunity

In the 1980s, the transnational corporations began to deemphasize operations in the Northern Forest. Today the old mills of this region cannot compete with the modern, more efficient mills of the southeastern United States that were built with record industry profits in the past decade, and the degraded pulpwood forests of this region cannot compete economically with the pine plantations that are beginning to produce marketable wood fiber in the southeastern states.

Ironically, this presents us with a once-in-a-lifetime opportunity to recover local control over the destiny of our communities and our regional economy. Several factors contribute to this opportunity. First, industry's long-term strategy of shifting the primary focus of its operations[26] to the Southeast and beyond the borders of the United States provides us with both the opportunity and the necessity of creating a new regional economy. It should be locally controlled, diverse, and based on ecologically benign, value-added manufacturing and sustainable community agriculture.

Second, huge amounts of industry-owned land are for sale today—estimates range from 3 to 5 million acres—and even more forest land will almost certainly be offered for sale in the next couple of decades. This affords us the opportunity to restore these forests to public ownership so that we can manage the lands for ecological and social values.

Third, the Northern Forest Lands Council, despite its many flaws, has provided a unique forum to analyze and discuss issues that are so critical to the fate of this region. In December 1992 the council took two important steps. A study commissioned by the Land Conversion Subcommittee of the council debunked the industry's claims that loss of favorable capital gains treatment, heavy property taxes, and onerous environmental regulations are forcing the larger owners to sell forest land. It found that most large land sales are due to such forces as changes in business strategy or a desire to achieve a higher return on investment elsewhere. Also, in December 1992 the council held a forum on the status of biological diversity in the region at which even the industry biologist on the panel, Sharon Haines of International Paper, agreed on the need to establish a system of ecological reserves in the Northern Forest region.

Finally, the resiliency of the region's ecosystems—symbolized by the return of moose and beaver—permits us to dream this dream of recovery. A sustainable human economy can only be possible if there is a healthy, dynamic forest. The Northern Appalachians are resilient due to abundant precipitation and adequate soils. But we cannot count indefinitely upon their resiliency. If we permit current abuses to go unchecked for much longer, ecological recovery will require millennia, not decades and centuries.

We live in a world of physical and ecological limits, whether or not we acknowledge them. If humans are to survive, we must develop a culture that provides for fundamental human needs while respecting natural limits. We should view this ultimatum as an evolutionary challenge to our species' survival. Our continued refusal to live within natural limits will spell our doom and the doom of countless millions of innocent species.

As Donella Meadows has written, "That is our choice—to set our own limits rather than have the planet set them for us. A good way to start would be by questioning our vague, sacrosanct goal of growth. Growth of what? For whom? For how long? At what cost? Paid by whom? Paid when?"[27]

A Community-Oriented Transition Strategy

A vision is essential, but without a realistic transition strategy to get us from here to there, it is meaningless. If we ever arrive at the point where we as a species live truly sustainably, we will not need regulations, wilderness reserves, buffer zones, and multiple-use forests any more than wolves, bears, or mushrooms require such designations. But in the interim, the long, difficult process of reconciling ourselves with natural communities will require these and other formal designations.

We must adopt several complementary strategies concurrently that address community, economic, and ecological issues. Here is a partial listing of the most important elements of a transition strategy:

1. Community empowerment of those who have been disenfranchised in the past offers the only hope of galvanizing broad-based community support for ambitious preservation strategies. In the past, communities in this region passively waited while industry, government, and out-of-region environmental groups determined their fates. Today, the citizens of the region realize that the absentee owners have betrayed them, and there is growing

support for local self-control initiatives. Some community empowerment groups that include a broad cross section of the community (mill workers, teachers, business people, clergy, senior citizens, artists, farmers, and environmentalists) have already been convened. Despite many differences, we all love this region and we cherish the natural community as much as the human community.

2. We must identify underutilized community skills, address the needs and interests of the entire community, and create a unifying vehicle for the community to regain local self-determination and confidence.

One strategy to promote cultural restoration is the creation of an academy that teaches ecological restoration and socially and ecologically responsible economics and business practices, and provides vocational training in traditional arts and crafts. A natural history museum with small satellite interactive museums that celebrate the diverse aspects of regional culture must be aligned with this academy.

3. We must reform the process of land development. One of the threats to the region comes from the development of lake, pond, and riverfront areas, land with scenic views, and land near developing areas such as ski areas. The timber industry, the Northern Forest Lands Study and Council, and some mainstream environmental groups have attempted to portray second home development as the most serious threat to the Northern Appalachians. While it is true that certain areas, notably the Adirondack Park, and some of the developable lands in the other three study states are threatened by such development, most of the industrial forest is threatened more by current forestry practices. Thwarting second home development will do little to protect the ecological integrity of the industrial forest.

Nevertheless, development is a serious threat that must be addressed. Three specific reforms will deal with development pressures: (a) an immediate moratorium on development within the Northern Forest Land Study area, (b) adoption of Existing Use Zoning for the region,[28] and (c) a land-gains tax that penalizes land speculation.

Easements in which the development rights to a parcel of land are purchased or donated to the public or to private conservation groups are a favored strategy by many. However, easements usually cost 40–90% of the full value of the land. On land most threatened by development the costs will be highest—often 75–90% of the full fee value. Most easements negotiated in recent years permit the landowner to continue to practice clearcutting, herbicide applications, and other unsustainable forestry practices and are very expensive.[29]

Easements can play an important role as part of an overall strategy, but they cannot, by themselves, protect even a portion of the region's ecological integrity.

4. Economic sustainability cannot be based upon ecological abuse. Propping up our region's forest based economy with subsidies and technological fixes, as proposed by industry, is bad science, bad economics, and bad ethics.

We need to assess the nature, structure, and condition of the existing regional economy, including the economic and environmental impacts of the economy's responsiveness to global market conditions. This would include both short- and long-term trends, current and potential utilization of resources, and municipal, individual, and industrial energy efficiency.

Particular emphasis should be focused upon the role of the forest products industry in the region, including the amount of wood fiber consumed by each sector of the industry, products produced, exports of raw and finished materials, workers' compensation, unemployment, bankruptcies, and the economic costs of environmental degradation (e.g., the health costs of air and water pollution). This study should also assess the tourism industry, agriculture, and other economic activities that are important in this region.

We must transform current practices associated with the forest products industry. There must be an immediate ban on the export of raw logs (and timber-related jobs). Loggers must be paid an hourly wage, not a piece-rate wage. Worker safety must no longer be sacrificed to industry profits. We must immediately ban the chlorine bleaching process entirely. Shifting from chlorine gas to chlorine dioxide is not a solution because it still produces hundreds of toxic organochlorine compounds in addition to dioxin (2,3,7,8-TCDD). We must examine the end uses of paper and wood products, and cease the manufacture of wasteful products such as paper towels, packaging, and junk mail.[30] This will significantly reduce the demand for virgin pulp. We must convert to recycled paper production. This recycled paper must not be de-inked nor bleached by the chlorine process. Mill price fixers must be prosecuted. Tax cheating on intent-to-cut forms and mill stumpage reporting must be prosecuted. Clearcuts, shelterwood cuts, seed tree cuts, regeneration cuts, high grading, pesticide spraying, new road building, reliance on heavy harvesting machinery, and whole-tree harvesting must be ended.

By viewing the Northern Appalachian forest ecosystems as communities to which we belong, not commodities that we own, our economic and social activities will be oriented to survival, not accumulation of excess wealth; subsistence, not profit; production for use, not for a global market exchange. We must produce a diverse mix of small-scale forest products such as quality furniture, boats, and wooden toys. The economy must become steady-state, not growth-driven.[31]

Several economic studies are currently underway, but most assume that we need new capital, new smokestack industry, and more of the same industries and absentee capital that have ravaged the region's economy and ecosystems. These efforts assume that our economic strategy must become even more dependent on the global-market economy that is characterized by free trade agreements, transnational corporations, absentee ownership, and resource extraction.

We must distinguish between the price and the cost of a product. The cost includes such "externalities" as pollution costs, health damage, environmental degradation, and lost economic opportunities (such as the loss of opportunities for subsistence and recreational fishing below a dioxin-spewing paper mill). Our accounting system must be based upon the true cost of producing a product.

We need to assess economic alternatives that are ecologically benign, promote diversity and long-term economic stability, assure local control of the region's economy (thereby promoting independence and insulation from absentee global market forces), create jobs that promote worker dignity and health, and emphasize value-added, labor-intensive small businesses and cottage industries in the forest products sector. Attention should also be focused on ecological restoration, ecologically benign recreation, and tourism jobs that offer more than the traditional menial service jobs.

We need to explore the possibilities for developing alternative business opportunities such as furniture making, canoe building, toy manufacturing, local crafts, and agricultural products such as jams and jellies and woolen products. The jobs should be tailored to the interests, skills, and needs of the entire community. Currently, markets for local crafts, agricultural products, and value-added forest products are meager. We need to develop North Country marketing cooperatives that can help local producers reach urban, suburban, and tourist markets in the region.

Ecological restoration will provide an immediate source of jobs. The creation of a new Civilian Conservation Corps would provide ecological instruction and job training, and do useful restoration work. It should be designed to "graduate" citizens who have marketable skills in ecologically sustainable work.[32]

Restoring Ecological Integrity

To assure the ecological and evolutionary integrity of the Northern Appalachians we must preserve and expand the wild habitat in the region and must take all necessary steps to restore extirpated subspecies and species and damaged communities and ecosystems. The preservation of biological diversity requires the preservation of large, undisturbed ecosystems—

more broadly thought of as wilderness. Several aspects of natural forest ecosystems in the Northern Appalachians are of particular importance:

Old growth. There has been an enormous outcry over the destruction of the forests of the tropics, where 30–50% of the old-growth forests have been cut down. Yet in the eastern United States, we have eliminated over 99% of the old-growth forests in the Appalachians. Fewer than 7,000 acres of old growth remain in Maine, a state of over 20 million acres. The destruction of eastern old-growth forests has been so complete that we honestly don't know what we have lost. We can list some of the larger, more charismatic species like bison, elk, and wolverine, but we have no way of knowing what we have lost of the more enigmatic microflora and fauna, nor the complex relationships and processes that have been disrupted. We can, however, gain a sense of the losses by studying the remnant old-growth forests of the northwestern United States.

Old-growth forests are usually composed of uneven-aged stands of dominant and subdominant tree species that create a layered appearance with old trees towering over younger trees. Snags, broken tree tops, and fallen logs litter the forest floor. The larger habitat space created by bigger trees is essential to many endangered species of birds and arboreal mammals. Scientists now suspect that old forests are required to regenerate young, healthy forests because they support cavity-nesting birds that eat damaging insects,[33] are critical for populations of invertebrates, fungi, and microorganisms, and play a central role in many ecosystem functions, such as water control, climate regulation, and nutrient cycling.[34]

Soils. Thoreau wrote of the Maine Woods: "The surface of the ground in the Maine woods is everywhere spongy and saturated with moisture. I noticed that the plants which cover the forest floor there are such as are commonly confined to swamps with us. . . ."[35]

Two years after Thoreau's death, George Perkins Marsh observed that after the lumberman has let light into the swamp and the forest disappeared, "all is changed. . . . The climate becomes excessive, and the soil is alternately parched by the fervors of summer and seared by the rigors of winter. . . . The face of the earth is no longer a sponge, but a dust heap. . . . The earth stripped of its vegetable glebe, grows less and less productive."[36]

Wildlife. In a healthy ecosystem, species at all trophic levels, from primary producers to decomposers, fulfill their independent and interdependent roles. Several groups of animals, however, are of particular concern in the Northern Forest. Amphibians, for example, are of concern here because of their position in food webs in forest ecosystems, and the fact that they are in sharp decline globally.[37] The biomass of salamanders per acre of forest is normally greater than that of any other vertebrate, including deer and moose. Clearcutting dries out salamander habitat and eliminates dead, rotten logs. Failure to regulate clearcuts and protect headwater streams and vernal pools is destroying the breeding grounds for amphibians, and their populations are further fragmented and isolated by roads.

There has also been an alarming decline in the populations of neotropical migrant songbirds in the past decade. Nationally, the greatest declines have occurred in Vermont, with New Hampshire and Maine showing the second and third worst declines. Two of these declining species, bay-breasted and Cape May warblers, are major predators of the spruce budworm. Recent studies indicate that these declines are due in significant measure to habitat destruction and fragmentation in eastern North America. It is important to note that the decline of songbirds in New York State, with 2.4 million acres of designated Wilderness and Wild Forest in the Adirondack Park, has been much less dramatic than in northern New England.

The needs of large, endangered, and extirpated native carnivores merit special attention in the Northern Appalachians. Predator populations are an indicator of ecosystem health. They are very sensitive to anthropogenic disturbances and have often been the target of extermination programs such as are currently waged by the states of the region against the eastern coyote. The plight of gray wolf, wolverine, cougar, and lynx in the northern Appalachians demonstrates that existing reserves are inadequate to sustain viable populations of these natives.

A new reserve strategy. If we are to achieve our goal of sustainable natural and human communities, we must take all steps necessary to allow for the natural recovery of damaged ecosystems and threatened, endangered, and extirpated species and communities, particularly the establishment of a system of ecological reserves. Conservation biologist Dr. Reed Noss believes the most important considerations for designing and managing ecological reserves that will maintain the native biological diversity of a region "in perpetuity" are to:

1. Represent, in a system of protected areas, all native ecosystem types and seral stages across their natural range of variation.

2. Maintain viable populations of all native species in natural patterns of abundance, distribution, and genetic diversity.

3. Maintain ecological and evolutionary processes, such as disturbance regimes, hydrological processes, nutrient cycles, and biotic interactions, including predation.

4. Design and manage the system to be responsive to short-term and long-term environmental change and to maintain the evolutionary potential of lineages.[38]

To achieve these goals we must establish large contiguous wilderness core areas—starting with existing National Forests and other public lands—connected by wide, wild corridors buffered from human disturbance. It is unreasonable to believe that this strategy will be successful unless at least 50% of the entire region is incorporated into core ecological reserves.[39]

Further, such a strategy must be regional. Although the Northern Forest Lands Study area provides an important starting point, we must extend our vision to include all of the White and Green Mountain National Forests, as well as the Berkshire and Taconic Mountains. Eventually, northern Appalachian reserves must connect with reserves in the central and southern Appalachians, the Midwest, eastern Canada, and the Maritimes.[40] Connectivity with other regions, especially on a north–south axis, such as the Appalachian Trail, is critical to mitigate predicted effects of global climate change.[41]

Conservation easements can be useful tools in protecting the integrity of buffer zones and protecting the rights of landowners of small- and medium-sized parcels in the region. If buffer zones are managed in an ecologically responsible manner, they will make the core areas more effective in maintaining a region's ecological integrity. Conversely, if areas near reserves are managed in an ecologically unsustainable manner, core reserves will need to be significantly larger. It is imperative, therefore, that we manage public and private lands in a complementary fashion.

Restoration. Because of the resiliency of the region, and the extraordinary opportunity to purchase large tracts of industrially owned forest land in the near future, the prospects for the natural restoration of biological diversity are promising. Whenever possible, restoration should be left to natural processes.

However, there will be a great deal of work for human restorationists, especially in the early days of this transition period. Closure of roads, erosion control, toxic cleanups, and the removal of exotic species are some of the most critical jobs.

Restoration work provides an ideal opportunity for the creation of new jobs to replace the jobs lost in the woods and mills due to industry flight. This healing work will provide the restorer an opportunity to learn to live in harmony with his or her home region.

Wilderness for its own sake. There are numerous practical reasons for defending wilderness and for demanding the restoration of vast tracts of wild land. Four billion years of evolving life have flowed from wilderness. A species cannot survive if it destroys its environment. Other species have rights, and they need wilderness. Wilderness contains cures to what ails us, from undiscovered medicines to the solitude that soothes the weary refugee from the folly of modern civilization.

While these reasons may be valid justifications for protecting wildness, in the end they are irrelevant. We do not debate whether or not to preserve the sun or the moon. We accept and celebrate their existence. And so it is with wilderness. There is no need to justify 4 billion years of evolving life. We are brief visitors to this unimaginably beautiful pageant of rocks and ice and fire and water and dancing sunlight. We have been blessed beyond fathoming. The greatest, indeed the only, necessary justification for respecting wilderness is that it is. Wilderness exists for its own sake, not ours.

But for our sake, and for the sake of the unborn, wilderness restoration is the key to restoring sustainable natural and human communities in the northern Appalachians.

Land Acquisition

The current system of private ownership of the vast bulk of the northern Appalachians has failed to protect biological diversity. Minor tinkering with this system cannot restore ecological integrity. Small, isolated reserves will not halt the loss of genetic and biological diversity. Even in the Adirondack Park, where 42% of the nearly 6-million-acre park is off limits to logging and development, much more land must be managed as wilderness if the ecological integrity of the park is to be protected.[42] Only about 7% of northern New England is publicly owned, and very little of that is designated Wilderness. Furthermore, these Wilderness areas are far too small, too isolated, and not representative of all ecosystem types and seral stages of this region. And they are not effectively buffered from clearcutting, development, and other anthropogenic threats.

To assure sustainable natural communities, we must purchase 10–15 million acres of industrial forest land to be incorporated into a network of evolutionary reserves. Currently, 3–5 million acres in the 26-million-acre Northern Forest area are for sale, and additional millions of acres of industrial lands are considered "nonstrategic."

Nonstrategic lands are generally those far from a company's mills. In the wake of the hostile takeovers in the paper industry in the early 1980s, some companies separated their landholdings from other assets. The new land companies are wholly owned by the parent paper company, but they must now show a much greater return on investment than was previously the case. Hence, there is pressure on the land managers to inventory their holdings to identify the lands with development potential, lands that are not profitable, and nonstrategic lands far from mills. James River has stated its willingness to sell much of its land to the public.

An example of nonstrategic lands is the case of Champion International, which has paper mills in eastern Maine (Bucksport) and western New York (Deferiet). In October 1992 Champion announced that it would like to sell approximately 95,000 of its 145,000 acres in the Adirondack Park because those lands are predominantly hardwood, and its mill in Deferiet needs softwood fiber.

Champion also owns about 330,000 acres in the Northeast Kingdom of Vermont and Coös County, New Hampshire. These holdings are far from Champion's mills, and therefore are nonstrategic. It has long been a poorly kept secret that Champion would be happy to sell these lands; in October 1992, a representative of Champion told some Vermont timber owners that it would be interested in selling some of its holdings in that state.

International Paper, Bowater, and other large landowners also have significant holdings of nonstrategic lands. Some employees of these large paper companies would like to sell these lands. Environmentalists would love to buy them before they are stripped of trees.

Although large tracts of land have been for sale in this region for several years, the major private buyers are not leaping at the opportunity. At a forum sponsored by the Land Conversion Subcommittee of the Northern Forest Lands Council, a panel of experts on the forest products industry explained why.

Evadna Lynn, First Vice-President of Dean Witter Reynolds, Inc., and a leading Wall Street analyst of the forest products industry, stated that corporate landowners are selling land either because they have a cash problem, or because they recognize an opportunity to liquidate an asset whose value has, until recently, been illiquid. In other words, they are taking advantage of a recent rise in the value of timberland. But, she added, "the Northern Forest is being bypassed by this resurgence in sawtimber values. As a rule of thumb, the northern commercial forests have a value around $100 an acre, compared with $400 an acre in the South and $1,000 an acre in the Pacific Northwest. These differences are based on stand density, regeneration cycles, and the ratio of pulpwood to sawtimber. Few paper companies appear to be looking to their northern timberlands as a source of cash, again seeing them more as insurance against shortages of fiber."[43]

Another panelist, Richard Smith of John Hancock Pension Fund, told the forum that institutional investors such as Hancock had recently begun buying huge tracts of timberland. In 1985 Hancock held only $70 million in timberland; today it has holdings valued at $1.2 billion. Overall, institutional investors currently own $2 billion in timberland, and Smith projected that this figure would rise to $6 billion in the next couple of decades.

Most of the $2 billion is invested in the South, especially in the Piedmont Range. There is about $750 million invested in the Pacific Northwest. Hancock owns about $15–20 million worth of land in the Northern Forest region, including almost 300,000 acres that it recently purchased from Diamond Occidental/James River for about $100 million. When asked why institutional investors were not buying more of the 3–5 million acres currently for sale in this region, Smith replied that the "economics" weren't favorable. When pressed, he admitted that given the condition of the land—massive clearcuts and the preponderance of mostly pulpwood and chipwood—the asking price for timberland in the Northeast was too high. In other words, abusive, unsustainable forest practices do affect land conversion strategies, and they have scared away the most enthusiastic new private investors.

This lack of interest by the private sector affords the public an extraordinary opportunity to acquire large tracts of the Northern Forest for the protection of public values such as biodiversity, water quality, public access, and a reliable source of well-managed working forests.[44] When environmentalists, affected communities, and the paper corporations get together to advocate public acquisition of these lands, the "impossible" could happen quite quickly.

The price per acre varies, depending on many factors. Prime development land is most expensive, but there is relatively little of it, and its development value could be reduced by existing use zoning. Large tracts of nondevelopable land could go for $80 to $200 an acre. The October 1991 sale of 2.1 million acres of Maine by Georgia-Pacific cost Bowater $155 an acre, if we assume that three mills and the hydroelectric rights to the West Branch of the Penobscot River were thrown in absolutely free. Even if we assume that an average acre will cost $150–200, it would cost less than $2 billion to buy 10 million acres. When we consider what our government routinely spends money on, we see this as a veritable steal.

These lands will never be cheaper than they are today, and the ecological integrity of these lands will almost certainly continue to decline under current private ownership, especially ownership by absentee transnationals. Therefore we must secure adequate funding to purchase the lands that are for sale or that will be for sale in the next decade or so.

Federal acquisition is clearly the only feasible way to purchase such large quantities of land. In these times of budget crisis and mounting deficits, it appears, at first glance, to be political folly to advocate more governmental expenditures. Yet despite the budget crisis provoked by the 1980s economic policies, Congress somehow always manages to find $500 billion or so to bail out the scandal-plagued savings and loan industry. There is always money for absurd military hardware. We can always afford more subsidies for the transnational corporations and the wealthy. The message is that money is available, but it generally is spent on death, destruction, and greed. We need a radical transformation of values.

I propose that a relatively small amount of money be spent in the northern Appalachians for life, for sustainable natural and human communities, and for assurance that regional economies will be able to meet the future needs of the region's population in an ecologically sustainable way. This modest investment will repay our economy many times over because it will become the basis of a sustainable regional economy.

Funding sources include increasing funding for the Land and Water Conservation Fund and redirecting misallocated money that currently is spent on military hardware, savings and loan bailouts, subsidies to agribusiness, and subsidies to the interstate highway lobby. Other funding strategies include tax breaks for the sale or donation of land to the public or to private conservation groups. If we put as much energy into funding land acquisition as we have put into fighting the Cold War or sheltering money from the IRS, we should have no problem developing creative new ways of raising adequate sums.

There is strong public support for the acquisition of land by the public, despite noisy campaigns against acquisition by special economic interests and their political allies. Opinion surveys in the Northern Forest area in the past several years have consistently found that 75–85% of the residents of the Northern Forest region support public acquisition of these lands to protect them for public access, to protect forest health, and to assure local control. In 1990 the Northern Forest Lands Study reported that a survey carried out in the Northeast Kingdom of Vermont and Coös County, New Hampshire found that "85 percent favored public acquisition for wilderness protection, 80 percent favored public acquisition to maintain recreation opportunities, 81 percent favored acquisition to maintain wildlife habitats and 72 percent favored public purchase to assure a future timber supply."[45] Unfortunately, the absentee owners, the real estate industry, private property rights advocates, and other special interests that benefit economically from the status quo have maintained the fiction that there is no support for public acquisition.

In September 1992 the citizens of Granby, Vermont, voted unanimously to raise $55,000 to buy fee title to 1639 acres of land surrounding Cow Mountain Pond from Champion International. Additional funding came from the new federal Forest Legacy Program, the Vermont Housing and Conservation Board, and the Connecticut River Partnership Program. Granby is a small rural town that relies almost exclusively on the forest products industry for jobs. It is exactly the sort of town that the opponents of land acquisition point to when claiming there is no support for public acquisition in the Northern Forest region.[46]

There is also growing support for a new public lands and economic recovery act for the Northern Forest region, a kind of new "Weeks Act." The 1911 Weeks Act, which created the national forests east of the Mississippi, was a response to the crisis provoked by the practices of the timber industry at the end of the nineteenth century. It is one of the great legacies of its generation. But after 80 years, we face a new set of crises that the Weeks Act is unable to address adequately. We urgently need a new act that recognizes that economic and environmental health are mutually dependent.

The economic recovery component of the act must provide assistance in transforming the Northern Forest economy to a regionally controlled economy that complements rather than exploits the natural world. It should provide assistance to small-scale entrepreneurs who wish to start up or expand existing value-added manufacturing or create marketing cooperatives for Northern Forest products. There must be support for reestablishing local agriculture for both subsistence and market, as well as support for comprehensive recycling, energy conservation, ecological restoration, and a Civilian Conservation Corps specially tailored to meet the needs of the region's youth.

The most important aspect of such an initiative would be funding to buy the millions of acres currently or soon to be for sale. Public ownership of large tracts of land is essential to protect the region's ecological integrity, to assure public access, and to complement responsibly managed private holdings. It is the key to the recovery of regional and local control over our economy.

A land acquisition strategy without an economic component will face a long uphill battle in Congress. An economic reform package without a visionary land acquisition component will fail to protect the region's environment and thereby fail in the goal of long-term economic recovery and stability. Only a package that combines programs for economic and ecological recovery can address the needs of the region's communities now and in the future. The price tag? Three to four billion dollars appropriated over a 5- to 10-year period. One billion dollars should be appropriated in the first year so that we can properly launch both economic and acquisition programs. Five hundred million dollars a year thereafter until the end of the century should provide adequate funding to promote sustainable natural and human communities in the Northern Forest. When we consider that $7 billion has gone to rebuild southern Florida in the wake of Hurricane Andrew in the fall of 1992, we see that a precedent exists for regional recovery packages.

Is this a radical, crackpot idea? Certainly the defenders of the status quo would have you believe it is. But their ilk also dismissed the public lands movement that created the eastern national forests through the 1911

Weeks Act. They dismissed Benton MacKaye's 1921 proposal to create an Appalachian Trail from Maine to Georgia as the idea of a fool. Yet the Appalachian Trail, one of the most treasured landmarks in the east, was completed less than two decades after it was first proposed!

Is it crazy or impractical to advocate ecologically sustainable and responsible strategies? Does wisdom dictate that we continue to pursue ecologically and economically disastrous strategies such as those that currently govern our Northern Forest region? If we develop a locally controlled economy that is ecologically sustainable, we will remove the economic cause of environmental degradation and we will empower our friends, neighbors, and community. Such a community will not need to be "educated" about the value of a healthy, vibrant environment. It will demand it. When the time comes for asking our political leaders to help buy land and develop healthy local economies, the absentee owners and their local agents will be unable to "divide and conquer" as they have for two centuries.

12

A Multi-Use Working Forest

Henry Swan

I believe that the future of the Northern Forest is as a multi-use working forest, not simply an ecological, environmental playground. My perspective in making this prediction is that of a private owner of nonindustrial forest land and manager of nearly one-quarter of a million acres of forest land in the Northeast. Wagner Woodlands is a timberland investment and forest management organization. We sell our logs and forest crops to sawmills and other users of wood. Our clients look at these activities as a source of return on their long-term investment in forest land.

Our income depends upon being able to harvest and sell forest products on a sustainable basis. Our growth depends upon our ability to find and acquire additional timberland for our investment clients. Frankly, we have substantially more money available for investment than we can find forest properties to buy. We are long-term investors and managers whose future depends upon a viable multi-use working forest, and we do not believe that a viable forest economy consists of simply growing trees to be cut down.

I believe that preservationists, conservationists, and forest industrialists can work together to achieve compatible goals. In the Northern Forest region we have a complex system, both economically and ecologically, that can potentially work to achieve everyone's goals, although currently it needs help to achieve this potential. In outlining my vision for the future of the Northern Forest and what we must do to achieve it, I will focus primarily on economic and social factors because they are part of the equation that is most misunderstood and where cooperation among all parties involved in the debate is necessary for the system to work. I believe that this debate is important; it is urgent that we take action now to halt the environmental changes that are being brought about by our near-term oriented society.

I firmly believe that a sound, long-term forest economy can exist in this region that also promotes biodiversity and protects forest values for recreation, wildlife, soil, and water. A working forest is important to the greater human community in a long-term sense. The main reason for this is people. Not just people who need to see extirpated species restored to the region, but also the North Country hardware store owner and the fly-fishing flatlander. Our forest is not solely a playground, although in my view forests provide a place for spiritual renewal and relief from the stresses of modern society. Personally, I get my renewal from my enjoyment of the forest. My main pleasure is fly fishing for trout and salmon (perhaps from a canoe on a river so I do not have to mind my backcast!), although I like nothing better than a challenging downhill ski in a foot of deep powder. I enjoy cross-country skiing with a group of friends, carrying a pack containing a mountain stove, chili, and a bottle of wine. My values are deeply rooted in the forest. However, I believe that these values are compatible with a multiuse forest, one that supports a healthy forest economy and still promotes its spiritual and ecological values.

A Working Forest

The term "working forest" is overworked. I would like to avoid this phrase, but I have not yet figured out a better one. To start, let me say that I believe that every acre of land has a greater use than to just hold the world together. Every acre of forest, as well, has a job to do, often many jobs. In my business of timberland investment and forest management, I like to see as many of the acres in our ownership as possible generating income. We need to achieve acceptable returns on investment that are competitive with other investment vehicles. Income for our company comes from the long-term production of forest crops and the continued capability of the land to do so, although we also gratefully accept income from recreational leases.

When I look at a working forest, I see two important values. First and foremost, the forest is a source of livelihood and income for the people who live and work around it. This is a direct economic value. Second, it has an ecological and open space value. Every acre of forest does not have to generate income from timber harvesting. We also need wild and undisturbed areas, and there are many acres of forest land better suited for that than for timber production. Those acres are working whether my forest industry friends like it or not. They provide habitat for animals and plants that cannot prosper in areas dedicated to cutting timber, and they provide an unspoiled natural setting for our children and their children to see "what it used to be like." Some of the most spectacular of these areas are not commercially viable for forest production anyhow. Some are old-growth stands that must be preserved. Hopefully everyone understands the need for watershed protection, and the tremendous ecological significance of bogs, swamps, and wetlands.

The Economics of the Working Forest

Financially the Northern Forest is on thin ice, and profit margins are weak at best. For many reasons, much of our commercial forest land does not now have the quality or stock of timber to provide an acceptable return on investment at the owner's expectation of market values for timberland. Why is this? First and foremost, because of a short-term stock market orientation toward current earnings per share, many public companies have been and still are forced to overcut their lands to generate income. In past years when favorable stock prices were not maintained, a series of takeover artists, like Sir James Goldsmith and Maxxam, Inc., recognized that the sum of the parts was worth more than the whole, and then proceeded to dismember numerous companies. Timberlands were either overcut or sold or both. Here in the Northern Forest, Goldsmith acquired Diamond International Corporation in a hostile raid on their stock. He ended up selling the mills for a sum large enough to recover his investment and was left owning hundreds of thousands of acres of timberland at virtually no cost. In the western United States, Maxxam, Inc., borrowed enough money through the sale of "junk" bonds to buy conservatively managed Pacific Lumber Co., sold off some fine diversified businesses and real estate, and then began to clearcut high-quality redwood stands to enhance its fortune.

The problems created by the takeover of Diamond International still exist in the Northern Forest. Goldsmith sold tracts of land in New York and New Hampshire to buyers who subsequently went bankrupt. A successor company is still trying to sell large acreages in Maine. Two other large forest products companies in the Northern Forest took steps to reduce the risk of a hostile takeover that have required them to generate more annual income than the forests can sustain. The pressure to generate current income comes from other sources as well. On private lands, as generations of landowners multiplied, too many beneficiaries try to derive support from too few acres. High real estate and inheritance taxes create financial pressures as well. These latter situations particularly lead to the practice of high grading, which involves cutting the best timber and leaving the rest behind.

We need to preserve our timber base for a healthy forest economy. We must have a sustainable long-term supply of quality timber. One problem with achieving this is land fragmentation: As land is split up into smaller units, much of the land goes out of production largely because the sizes of the tracts are not economically significant to the logger or landowner. Land fragmentation is not new, but I think we are more aware of it today following the activities of Patten Corporation and other irresponsible developers.

Patten Corporation pursued a strategy of buying large tracts of timberland at wholesale prices—about $150 per acre—and selling them as separate, smaller parcels at retail prices many times that—sometimes $2,500 per acre or more. To achieve these retail prices, they did intensive subdivision, mostly on paper. The value they added to the properties included laying out and perhaps rough grading the roads for the subdivisions, conducting enough soil tests to certify that individual lots could support septic systems, and developing an aggressive financing program for buyers. Because some towns had weak subdivision regulations and state requirements for septic systems had loopholes, Patten usually chose towns with no zoning regulations and then circumvented septic system requirements. With skillful lawyers and experienced land agents, they easily overwhelmed weak local governments. However, their sales practices eventually brought injunctions from the Attorneys General of a number of northern states. Unfortunately, many rural towns, particularly with lake and river frontage, have been left with the remnants of Patten subdivisions. In addition, many buyers, thinking that they had acquired an ideal lot for retirement, investment, or recreation, have been left disillusioned, in many cases with unbuildable lots and unfinished subdivisions.

Another problem is the overcutting of existing forest lands. If the purchase price warrants, some tracts will be bought, clearcut or high graded, and resold as smaller lots. It is clear that if the timber base shrinks, it will be more difficult to attract permanent long-term capital for forest industry investments. Because of our region's good growing conditions, geographic location, and tree species, we are well positioned to become the premier source of quality solid wood in the world, but the Northern Forest region needs new investment capital for the industry to be competitive in the world marketplace.

Possibly we have reached a point of turnaround for the forest industry in the Northern Forest. I think that we have at least bottomed out. The worldwide shortage of quality solid wood has allowed the forests in this region, which are now maturing after the first round of exploitation from 1850 to 1910, to again become an important supplier. Our ability to fill this need is based on several factors. First, we are seeing succession to a climax forest in most areas, hardwoods or softwoods, depending on the site. Fortunately, most of the trees in these climax forests are of economically important species. Also, because of the overexploitation of tropical hardwoods, the global market has had to turn to our finer-grained northern hardwoods, such as cherry, maple, and birch. Worldwide demand continues to be high for open-grained oak and ash. Finally, overharvesting of trees on private lands in the Pacific Northwest, along with the removal of public lands from timber production for environmental reasons, has also led to a scarcity of softwood construction materials.

The continued high demand from timber-starved Pacific Rim countries has led to their return to the Northern Forest for purchase of construction and millwork grades of softwood. Stumpage prices, the prices that timber owners receive for the standing trees in forests, are going up. We saw a doubling in spruce/fir stumpage prices in 1991–1992, and it continues to climb. In 1991, quality white pine stumpage doubled and now these price increases have spread to all grades of pine. Stumpage prices for hardwoods are breaking historical trend lines by sharply increasing. Depending upon species, real (inflation-adjusted) annual rates of price increases have gone from the historic 3 or 4% to 8% and more, even at a time when new housing-related construction is low.

I see little prospect that these trends will change in the near future. I have recently worked in two of the Russian Commonwealth of Independent States and have counseled the government of an eastern European republic. I doubt that any of these areas will threaten our timber industry in the near future.

I think that the Northern Forest Lands Council provides us a mechanism to help achieve a sound forest economy. To date, the council's greatest contributions have been to stimulate awareness and discussion of the issues. I believe that the Forest Stewardship and Forest Legacy Programs would not have become part of the 1990 Farm Bill had it not been for the efforts of the predecessor of the council, the Governors' Task Force for Northern Forest Lands, of which I was a member. Although I think the council has encouraged too much polarization on the issues, I believe that out of this controversy will emerge a workable middle ground. People are talking, and unique partnerships are developing. Yet, despite the positive indicators for the future of the Northern Forest, the threats to a multi-use working forest are real and need to be addressed.

Threats to a Multi-Use Working Forest

In analyzing any system, natural or political, the threats to it must be examined. I classify the threats to our multi-use working forest in three groups:

physical, societal, and economic. In assessing the physical threats today, we have to look at both those that are natural and those that are caused by humans. Fire, wind, ice, insects, and disease are the prime natural risks affecting our forests. Rather than going back in history and tracing how the operation of these have influenced the composition of the forests today, let me look instead to the future. Another outbreak of the spruce budworm will occur. The cycle of spruce budworm outbreaks has not been broken, although human activity has probably altered it. The current mixed-species composition of the Northern Forest, partially created by the last budworm outbreak and related timber-harvesting activities, may make future losses less serious. For both economic and environmental reasons, I do not think we will ever again undertake massive spray programs in an attempt to control forest pests. Moreover, we will be able to salvage affected timber and cut our losses.

We can look forward to periodic assaults of other insects and diseases. However, good forest managers have become more skillful at managing these problems. For example, Wagner Woodlands continues to buy, own, and manage oak lands on good sites even in the presence of the gypsy moth. We plant and manage white pine despite the existence of the white pine blister rust. We will have outbreaks of other significant pests such as the hemlock wooly adelgid and pear thrip. However, with an accessible forest, good surveillance, and reasonable management flexibility, we will cope.

Acid deposition and global warming are on the top of everyone's list as physical risks to the forest that are caused by humans. I do not think that their real effect or interaction with other systems is totally known. I do know we need to be more than just concerned bystanders. Years ago we noted some die-back in paper birch and were very concerned. One day one of our foresters took an increment coring and discovered that the trees had gone beyond their normal life span. The same situation occurred in one of our intensively managed, old sugar maple stands. I have other examples of selective timberstand mortality that some observers incorrectly blame on acid rain. The point is that stand mortality is a complex problem, one that is not fully understood and is perhaps easily mitigated. Although the potential threat of acid rain is real, it has not discouraged my organization from acquiring and managing Northern Forest lands, as long as they possess good quality sites for growing trees.

I wish we knew more about the effects of global warming and that our society was doing more about the probable causes. From a forester's perspective, I foresee the possibility of a northward migration of forest-type boundaries. However, because much of our forests is really in a species transitional zone, the commercial effects are off in the future. Yet I do worry about unique and ecologically significant alpine plant communities that are today at their southern limits in this region.

Timber-harvesting activities are also a potential risk to forests. For example, in a working forest we must be very careful not to stress through harvesting practices species that live on fragile sites. Many of the most fragile sites, because of inaccessibility, altitude, and forest stocking, are classified as noncommercial forest lands by responsible land managers anyway. On the other hand, foresters at Wagner Woodlands have noted that managed hardwoods on good growing sites are vigorous, while nearby sugarbushes of overtapped maples show signs of decline. My point is that humans can and do exert an influence on forest health on a site-specific basis.

Societal pressures create the greatest threat to the working forest today and will probably continue to do so tomorrow. When I think about societal pressures, I initially think of the usual: Expanding populations need more Northern Forest open space for recreational pursuits; communities within the Northern Forest need a balance between the outdoor recreation industry and the forest products industry; use of the forests, both for timber harvesting and recreation, potentially threatens biodiversity; and there are social problems relating to the property rights debate.

Presently, I am also greatly concerned about the misinformation being communicated by individuals and groups who further their objectives by irresponsible acts. On one hand, there is a vision that the Northern Forest should be a vast Wilderness Preserve and Playground. Some militant environmental groups believe that they have the right to tell landowners how they can use their land and what trees may be cut. To reinforce their position, they may resort to sabotage, such as putting in trees spikes aimed at damaging saws or damaging logging equipment by putting sugar in fuel tanks and sand in gearboxes.

At another extreme, we hear from some that current attention on the Northern Forest will lead to a major federal takeover of the area and that treasured hunting and fishing privileges will either be taken away or shared in an unacceptable manner. Some of those who use forest land for recreation believe that they have a right to use lands owned by others with minimal fee or restrictions because they have always done so. They focus their wrath on public agencies and conservation groups who they believe are spearheading a movement toward federal control of all Northern Forest lands and elimination of their free "entitlements."

The issue boils down to one of landowner rights. The two extremes I have just highlighted lose sight of the fact, for convenience I think, that someone else owns these forests, manages them, and pays taxes on them. It certainly does not help relationships between the landowner and the public when a gigantic Maine Woods Reserve is announced by the Wilderness Society, a leading national environmental organization, without at least some discussion with those who own the land.

On the other hand, militant private property advocates have themselves added to the polarization of discussions. Hearings have been interrupted, misinformation spread, and meetings terminated because the physical safety of the council members could not be assured against the threats of violence.

As a representative for substantial timberland owners and a participant in numerous Northern Forest task forces and committees for both conservation and forest industry initiatives, I have often found myself on the line facing fire from representatives of all these factions. Simply put, the process as it now stands is insufficient for the task at hand. Time is too precious to have to deal with all the conflicts and fears that impede honest interaction. Well-intended conservationists may very well be dealing with suspicious, unreceptive, and uncooperative landowners. On the other hand, with all this interest and debate, landowners may have a misperception of the value of their forests. These conflicts make the job of constructive change difficult, and must be overcome.

The final threat I see to the Northern Forest is one of economics. Obviously, it costs money to own land. There are two costs to recognize: current expenses and revenue foregone. Northern Forest landowners, like the rest of us who own real estate, are under serious real estate tax burdens. True, Maine has a Tree Growth tax, New Hampshire and Vermont have "Current Use" assessment programs, and New York has a relatively unworkable plan usually known as section 480(a) assessments. Because of declining state and municipal revenues coupled with higher costs for governmental infrastructure, landowners are being looked at as a source of additional revenue. The thought is often that landowners must be wealthy because they own land. The term "land poor" is seldom heard in our state capitals, but it is a fact! Vermont funded only 80% of its Current Use budget in 1992; New Hampshire's entire Current Use program is always under legislative attack; Maine once decoupled forest fire protection from the Tree Growth tax base, thereby increasing per acre taxation; New York cannot manage to enact a program better than section 480(a) and also has problems in timberland assessment apportionment. Landowner and forest industry groups are continually fighting to retain what little tax advantage they have and, as a result, are not in a position to craft a more universally acceptable tax program.

The problem of revenue foregone primarily exists for more financially sophisticated landowners. For example, corporate financial officers or personal financial advisors make the observation that the market values a parcel of timberland at $250 per acre. After they pay real estate taxes, management costs, road maintenance, and other operating expenses from the income of timber harvests, there is little profit left. The current income yield is only 1.5 to 2%. A company in the forest products industry can get 15 to 20% incremental returns on investments in plant and equipment. Besides strategic timber reserves, their capital need not be tied up in timberlands. An individual can invest in equities with a current yield of 2 to 3% and have a long-term total return in the range of at least 13%. It takes a good and well-managed forest to accomplish this. The percentages I quote are from actual observations and provide a scenario relevant to the economic decisions of timberland ownership.

It is important to understand some of the forces that drive the economic picture I have sketched, because they affect whether we can make the working forest "work." First of all, the Northern Forest has a shrinking resource base. I hope this can be stopped. For environmental and recreational reasons, an increasing percentage of our public land is being removed from timber production. In the private sector, landowners who have needed to increase income have subdivided their land. This land fragmentation has subsequently created tracts too small to interest foresters or loggers. In addition, studies show that new landowners may not be motivated to have forestry operations on their newly acquired property. Even if they wanted to do so, often they cannot because the previous landowner sold the quality marketable timber before selling the land.

I am not proud of another reason for the shrinking resource base: overharvesting. Because of the need for current income, which translates into current earnings per share for a public corporation or just plain income for others, many forest stands have been cut too heavily. While this may have solved a short-term problem, in the longer term it reduces the supply of available timber. A shrinking resource base means that we cannot keep nor attract needed forest industry in our Northern Forest. *A value-added forest industry is critical to the economic well-being of the North Country.* Simply cutting low-grade fiber or fuel wood may help some people, but sound economics start with primary manufacturing into paper and lumber. Secondary conversion into specialized paper goods and solid wood products such as furniture or furniture parts, or milled hardwood lumber for export, may now only be a dream but is achievable in the long-term if we work at it.

Yet despite all these threats, I do not believe that the forest products industry should just throw in the towel and let the preservationists take over, simply because it would not be in the best interests of the people who live here and need to make a living from the forest. Forest industry is a needed part of the region's economic diversity. Moreover, our forests possess potentially valuable timber species that will be increasingly important in our global economy.

Initiatives Afoot

There are a number of initiatives and positive economic factors that should encourage long-term management of multi-use forests and, hopefully, a rekindling of spirit in the forest industry. Following are a few of the initiatives.

1. *The Forest Stewardship Program.* This was part of the 1990 U.S. Farm Bill and is designed to assist nonindustrial forest landowners with long-term management of their lands. For example, in New Hampshire alone the objective of the program is to encourage the owners of 300,000 acres of forest land to draw up private stewardship plans by 1995. In keeping with the landowners' objectives, recommendations will address such issues as wildlife habitat, water quality, soils, recreation, aesthetics, forest protection, and production of forest crops. Each state has formed a Stewardship Committee and is developing programs to communicate and inspire landowners to participate. With more forest land under some kind of management, quality timber should be produced that at some point, when stumpage prices go up, will be available to the forest industry. Unfortunately, after initial funding, the program has fallen to the federal budget-cutting ax. Whether it will be funded in the future remains to be seen.

2. *The Forest Legacy Program.* This provides an opportunity to protect open space and promote forest management. It was also part of the 1990 Farm Bill. The four northeastern states in the Northern Forest Lands Study Area plus the state of Washington have been designated to undertake pilot projects in a joint federal/state program for forest land protection. The ultimate goal of the program is to protect environmentally important private forest land threatened with conversion to nonforest uses. To help maintain the traditional uses and integrity of private forest land, conservation easements will be the prime protective mechanism. Each of the states has a Forest Legacy Committee to select timberland candidates and establish the program, which requires a Forest Stewardship Plan or other acceptable management plan. The implementation of this program in Vermont is a good example of how it can function. Under the leadership of the State Forester and with the help of in-state conservation organizations, environmentally significant forests were identified, landowners were contacted, meetings soliciting local input were held, forest resource and forest

industry infrastructures were studied, and the resources of the U.S. Forest Service were wisely used; now, negotiations for two top-ranking properties are underway or have been completed. I expect Vermont's approach to be the model for projects in the other three northeastern states.

3. *The Reforestation Tax Act of 1991.* This is making its way through the Senate Finance and House Ways and Means Committees. Of particular note is a partial restoration of the long-term capital gains treatment for all timber sales and allowance for the deduction of annual forest management expenses as permitted prior to the 1986 Act. In addition, there is an increase in reforestation tax credit and reduction in amortization years that is more significant in areas where tree planting is more the usual practice than it is here. The bill addresses and helps restore some of the necessary economic elements of timberland ownership that were removed by the sweeping Tax Reform Act of 1986.

4. *The Northern Forest Land Council.* The NFLC, through its subcommittees, is implementing many of the recommended strategies of the Governors' Task Force on Northern Lands. In my view, their most important efforts are in the areas of state and federal taxation of forest lands and revenues from timber harvests, estate taxation, and protection strategies for productive timberlands. I hope the council is the catalyst for making this working forest more economically viable.

5. *Noteworthy state initiatives.* The most significant recently is the passage by New Hampshire's legislature of a small amount of bond guarantees for the James River Corporation's Berlin paper mill. Given New Hampshire's fiscal situation, the amount of bonds guaranteed is small ($35 million) relative to the size of the mill, but the endorsement provided positive encouragement to the corporation.

Each of our northeastern states has had successful land conservation programs. Maine has the Land for Maine's Future program. New Hampshire had its Land Conservation Investment Program, whose intended life span of 5 years has now passed. Vermont has its Vermont Housing and Conservation Trust. New York has spent the last of the money made available by its 1986 Environmental Bond Act and failed to pass another Bond Act in 1990. I am particularly pleased with the parts of those programs that have used conservation easements to protect open space and reserved fee acquisition for only the most environmentally sensitive and valuable areas. Unfortunately, due to the general economic downturn, state budgets as well as voters' pocketbooks are seriously depleted and state-funded land protection programs have generally been cut or put on the back burner.

Where Do We Go From Here?

In order to achieve a viable working forest, the economics of forest ownership must be made more attractive. Many recent experiences suggest to me that some sectors of the public no longer take for granted that the Northern Forest will remain in stable ownership and be productive over the long term. The sale of Great Northern Paper in Maine to Georgia-Pacific and then to Bowater, and the Diamond International land sales in New Hampshire, Vermont, and New York have fortified the view of the potential volatility of land ownership in the Northern Forest region.

Bowater must raise capital to modernize the mills it acquired in Millinocket and East Millinocket, Maine. Logically, because of its vast landholdings, some of the lands may be sold. Obviously, lands that are farthest from the mill will be sold first. In addition, lands that have a more valuable use, such as those with lake or river frontage, will be easier to sell and net the most money.

The initial buyers of the Diamond International lands are in financial difficulty. In fact, the lands in New York that were not bought by the state have been on the market for a long time and have attracted no buyers because of the heavy cutting of the trees by previous owners. Scattered former Diamond International lands in New Hampshire and Vermont are either tied up in bankruptcy, or resale is again being attempted by the current owners after they have removed much of the timber. (Fortunately, the largest Diamond International tract in New Hampshire has become the Nash Stream Forest, a joint state of New Hampshire/U.S. Forest Service partnership that insures a "working forest.") Diamond International still has lands in Maine that it is slowly selling on a parcel-by-parcel basis.

I return now to the issue of the lack of depth in the market for forest products, particularly for low-quality wood and less valuable tree species. The reason I must address this is that the argument in favor of a working forest can be easily rebutted by the simple observation that if markets do not exist for low-quality wood, then the simple solution would be to take these lands out of production. I disagree. These lands provide a timber base and incentive for entrepreneurship in creating new wood-using industries. We are already seeing studies on the use of pressure-treated red maple for many applications that formerly used southern pine. Clean hardwood chips for making paper are currently being exported to the Pacific Rim from the southern United States, and that industry is now moving northward. Our timber base will eventually attract investment, and with investment will come jobs.

However, we need active incentive programs to attract and retain forest-based industry. The forests of the Northern Adirondacks are in a crisis situation right now. Most of the good hardwood timber has been high graded. Because spruce–fir markets were strong in 1991–1992, those species were seriously overcut. In 1991, we lost all but one of our remaining markets for low-grade logs there, and that mill can make better money by importing hardwood cants, squared timbers that look like railroad ties, from Canada and sawing them into pallet stock for its markets here.

Obtaining funds requires good people with good ideas that are well researched, and backed by a reasonable business plan. Because of the market gap I described, private-sector initiatives are being formulated, albeit slowly. For example, a company in mid-state New York, of which I am a director, made the decision to build two new dry kilns for drying squares and lumber. We qualified for no fewer than 15 economic development programs. After sifting through them all, we focused on three, and in the final analysis may settle on one that is targeted toward the energy-saving technology incorporated in our design.

In the Northern Forest region, there are real opportunities to tie increased employment to the forest-based resource. Along with tourism, the forests, properly managed as a renewable resource, can support a sustainable economic base. Designation of enterprise zones will enable businesses to obtain low-cost capital. Investment tax credits could make replacement of outdated equipment as well as expansion of existing mills attractive. Programs for borrowing at low interest rates could focus on facilities that promise to maintain a clean environment. Job-training programs can be funded. All of these initiatives take leadership and teamwork, and demonstrate an area where public/private partnerships might successfully work.

Along with attempting to make the income side of the equation more attractive for the forest industry, we can do something on the investment side of the equation as well. Through proper mechanisms we can reduce the amount of invested capital the landowner has locked up in the land. Conservation easements are the best device I know of to accomplish this. We can view land ownership as a bundle of rights. In addition to mineral, grazing, and water rights, other rights are particularly pertinent to the Northern Forest. These are development rights, open space rights, recreation rights, and visual or aesthetic rights, as well as timber rights. Federally funded programs like the Forest Legacy and Forest Stewardship Programs are good mechanisms to compensate landowners for certain of their rights acquired for the public good.

The public does not need to acquire all of these rights. Timberlands can remain productive. The beauty of a conservation easement is that it can be a flexible document tailored to both the needs of the landowner and the desired conservation initiatives of the public, but a source of funding for easements must be found. Possibly it could come from federal revenues received from leasing offshore gas and oil rights. The connection of using income from one part of the bundle of land rights to acquire other bundles of rights has the right political "ring" to it. Or perhaps the future compensation from hydropower relicensing on public waterways can be applied to an easement acquisition fund.

I also support the easement approach because the landowner continues to own the land and the federal government does not get further into the business of land ownership and management. Our Yankee heritage is one of states' rights, and that principle is not violated by this mechanism. The timberland owner is still a taxpayer. The land management record of the federal government is not very good. Management should be left in the hands of the landowner, but not without hooks. However, the hooks cannot be as deep as national conservation groups demand. Forest management and/or forest stewardship plans must be developed and followed. Additional incentives for public access and use of roads might be developed. Adherence to easement requirements might be vested in peer-group review such as a State Society of American Foresters Committee or a Regional Conservation Organization appointed by the administering State Forester. We will also need to find a way to compensate these oversight organizations for their efforts.

The most singularly important benefit of a healthy forest industry is the diversification of the economy in the North Country. We have in the Northern Forest, or in any forest, a truly renewable resource that can be managed and manipulated. Yet the United States today is shifting from a producing economy to a service economy. *This is a very dangerous trend, one that I cannot emphasize too strongly.* In our North Country economy we have a tremendous opportunity, through productivity increases and new investment in capital equipment, to employ more people, improve the quality of life, and increase incomes. I estimate for every person working in the woods harvesting timber, another three to five people are employed converting that wood into salable products, transporting and selling it, and providing the necessary support services such as equipment repair, fuel transport, and running that hardware store I mentioned earlier.

I think we are going to see more change in the Northern Forest. There is going to be ownership change simply because of changes in landowners' objectives. This strategic revaluation is going on now. I do not see knee-jerk reactions to ownership change as we saw with the Diamond International land sales in New Hampshire and New York. An example of these reactions was the New York Department of Environmental Conservation hastily entering into purchase negotiations for a portion of the Diamond International lands that had been heavily cut and threatened with subdivision. They probably overpaid for the lands or gave up too much in easement conditions, because they reacted before developing a thoughtful and coherent long-term strategy. The only reason they did not buy all of the Diamond International lands was that they ran out of money.

In New Hampshire, about half of the Diamond International acreage was purchased from the owner at a substantial premium over what he paid for the property. Conservationists felt this land would soon be developed. Here again, the purchase was made without fully examining the underlying fundamentals of the then-existing economy cycle. In hindsight, an overexpanded banking system led to the collapse of real estate investments and the owner eventually went bankrupt.

Available state and federal money at the time allowed these frantic public purchases to take place. With massive federal deficits and state economies in very poor shape, it will be a long time before substantial funds are again available for the discretionary purchases of the past. There is now a great change in the dynamics of financial speculation. We do not have the banking and economic outlook that fueled the past frenzy.

Primarily, I see more emphasis on the productive capability of timberlands. Owners who took their profits today at the expense of tomorrow may not be able to easily sell their lands. If they do decide to sell, they may have to accept a greatly reduced price in order to motivate an investment-oriented buyer.

I would like to interject a word of caution and advice here. Historic land issues in the United States have been played out first in the Northeast: the kind of legislation that created the Adirondack Park, the Weeks Act that created the eastern National Forests, the land-use patterns seen in our early New England villages, and the recent use of transferable development rights to concentrate development. Unlike other regions of the country, in the Northeast we have always been able to resolve our issues by sitting down and talking about them in the New England Town Meeting form of decision making. This still works. Old Yankee participatory decision-making processes should not be fragmented, and all who participate in the debate over the future of the Northern Forest should realize that polarization will lead to disaster.

We can retain and improve multi-use working forests only if we all work together. No single use should have superiority over all others in our Northern Forest. However, I strongly believe that responsible land managers can accomplish multiple-use objectives even while keeping the land in productive growing forests.

PART IV

Final Thoughts

We end this book with two chapters that step back and draw together many of the thoughts presented in the previous sections. The first is an essay by John Elder, Professor of English and Environmental Studies at Middlebury College. Elder, a resident of Vermont for nearly 20 years, uses a series of trips into the local forest in search of the wreckage of a plane as a springboard to reflect upon the relationships of humans and nature in New England. This essay on the "marriage" of culture and wilderness helps us to consider the Northern Forest within a larger context.

In the final chapter, we assess the contributions of these authors toward a common vision of the future of the Northern Forest. We examine what we have learned from these chapters, and ascertain the common ground among the authors, as well as the major points of disagreement. We conclude with a rough sketch of a future based on bringing the best of these ideas together, while at the same time offering our criticisms of points made that stand in the way of moving forward toward sustainable human and natural communities in the Northern Forest.

13

The Plane on South Mountain

John Elder

Our villages in north-central Vermont are wrapped and backed by trees. Unlike the ancient forests of the West—vast, pristine islands amid the tides of settlement—these are the patchy woods of local memory. The oldest residents of Addison County remember well when the rugged pitch between Bristol and Ripton was still farmed, and when loggers with axes and two-man saws shot white pines down to Route 116 through a steep chute on the western cliffs. But since World War II these slopes have been abandoned, allowing dense second-growth forests to circulate back around our human enterprises. The high ground has reverted to a providential wilderness. What does it mean to live in a place where the trees have come back—a region where, outside of a few cities like Burlington, the human world has been enfolded by the wild?

Last August I followed an abandoned logging road up into the Bristol Cliffs Wilderness Area, looking for the hulk of a World War II fighter plane that crashed into South Mountain on October 24, 1945. I had learned about this old catastrophe when my next-door neighbor showed me the clippings about it in her scrapbook. A Curtiss Helldiver with two Navy pilots aboard was returning to its base in Rhode Island after having taken part in a military air show in Burlington. But the clearance was low that afternoon, and the pilots somehow lost track of their altitude. My neighbor Audrey told me that when the stocky, single-prop plane sheared through the trees and met the ground an explosion rattled windows in the village of Bristol.

I wanted to see beech trees reaching through those broken wings, shuffling history back into the forest's fluttering deck. But there was no plane to be found on this summer afternoon, despite the fact that I'd gotten directions from two people in town who had seen it for themselves. One had said to turn straight up the main ridge from the outlet of North Pond, then walk due south for twenty minutes. The other said to strike a line at southwest 220 from the pond; I would find myself climbing up a cut to the mountain's highest bog. There, amid hemlocks and spruce, I would discover the wreck. Neither of these routes worked for me, though. Hardwoods and puckerbrush grew so thickly in that season that it was hard to follow any line, while the corrugated terrain left me uncertain where the actual ridge was.

I soon realized that my directions would never bring me to the plane, so I began simply to criss-cross the mountain's eastern face where it went down. Sometimes I would spot the bluish needles of a hemlock, spruce, or white pine and stumble toward them, looking down with every other step because of toppled trees and branches littering that wind-torn slope. My heart lifted whenever a patch of sunlight made a granite slab or a beech tree gleam like alloy. But I soon lost hope each time, perceiving that I was not, after all, at the scene of a disaster.

I walked for six hours, up hill and down, looking for this human incident with which to punctuate the narrative of nature. The missing plane was so much in my eyes that I registered little of what I actually saw. My hearing was sharpest to the buzz of small craft passing overhead. "They're leading me," I thought, or, more ambivalently, "What if they should crash at the exact spot?"

I never found the plane on that August hike. A broader circle through the woods, and through the seasons, was required before I reached the spot. One clear morning in early December of this year, our family's retriever, Maple, and I set out on a new expedition into Bristol Cliffs. It was a bright day, holding at around 30 degrees, and the roads in our village were bare. So I wore my running shoes in order to get up to the ridge faster, and to have more time before dark to comb that hummocky slope for the wreckage that eluded me on the previous attempt.

Just the couple of hundred feet in elevation between my house and the cliffs turned the ground white. At the trail's beginning, frost coated the twig-strewn track so heavily that it formed a frozen lattice. As I jogged uphill the interstices filled in, smoothing the grid into a blank new page. Soon after that I slowed to a walk, as the snow reached ankle deep. But Maple continued to career over logs, her reddish golden fur looking warm and alive against that glinting world.

When the last tracing of those ancient log trucks disappeared, I followed an outlet creek uphill. It trickled from the stump-surrounded stockage of a beaver dam. The beavers' pond, brought into focus by a thatchy, conical lodge, extended north–south in a little valley of its own. Maple and I slid along the ice until it led us to North Pond itself—a perfect oval set among steeply surrounding hills. Blueberries grew thickly along the western side, their tough leaves leathery and purple at this season.

I clambered up the slope, heading for the "heighth" of land where one man had told me the plane would lie. But, as Maple and I climbed up and down during the next two and a half hours, I realized once again how hard it was to tell which portion of the broken and interfolded range was actually the ridge.

I was just about to give up once more, so as to make it back down to my car before darkness came to this short day, when I spotted the hub of one of the plane's wheels. It rose against the gray-white background in an arc of patchy blue metal, its curve highlighted by a ring of ice within the hub. I scrambled down into a little swale to take a look at this wreck so little like what I had imagined. In my mind's eye, there had been a largely intact fuselage of silvery aluminum, with the wings still attached though perforated by straight young beeches. Actually, there were large hunks of green or blue metal scattered around a circle almost thirty yards in diameter. It looked as if a plane made of blue ice had struck a rock and shattered.

The biggest remnant was the engine block. Huge, finned cylinders rotated out from the block proper, with a massive shaft showing where the single propeller was attached. In addition to the engine and the single wheel, I found part of the tail assembly, as well as pieces of the wings and belly. Only the wing flaps remained totally undamaged. Curved sheets of metal perforated with circular holes an inch in diameter, they would have slid out to slow the plane when it went into a dive—delivering the thousand-pound bomb carried internally in a Helldiver.

Explosion and decay have obviously been helped out, in the demolition of this plane, by souvenir hunters. No insignia remained, though on one scrap I did see the single tip of a white star, shining from what was once the rim of a painted blue circle. The tail itself, the propellers, the windows, the landing gear, the control panel, and the seats have all been carried away. I've heard that two machine guns originally mounted on the wings were skidded down to North Pond and sunk out of harm's way. And I have to admit that I carried one relic away with me, too—a twisted piece of olive drab aluminum the size of a fallen leaf. It's here beside me as I write, on the table by my word processor.

It figures that I found the plane in early winter. The bewildering foliage of those hardwood groves had all rotated underfoot, and much of the brush had been knocked down by the early storms. In his entry for January in *A Sand County Almanac*, Aldo Leopold wrote,

The months of the year, from January up to June, are a geometric progression in the abundance of distractions. In January one may follow a skunk track, or search for bands on the chickadees, or see what young pines the deer have browsed, or what muskrat houses the mink have dug, with only an occasional and mild digression into other doings. January observation can be almost as simple and peaceful as snow, and almost as continuous as cold. There is time not only to see who has done what, but to speculate why.

Like all people who live in snowy regions, Vermonters are familiar with this process of reduction and focus. The cold months settle into our state as a gradual clarification. Winter holds up objects in high relief—boulders sealed in globes of ice, strawberry-colored blades of grass twisted through the frozen lacework at a pond's edge—for our most careful regard. It invites us to be still and cool, to let one curve, one color truly enter the mind.

Winter is just one of the erasures through which Vermont has come into its own. Leopold had originally planned to call his book *A Sauk County Almanac*, after the Wisconsin county where he worked at reclaiming a worn-out farm. But "Sand County" is inclusive of a broader locale—any of the districts across the United States where erosion, drought, deforestation, or just plain bad soil uprooted farming communities and replanted the fields with a second growth of trees. Much of northern New England and upstate New York belongs in Sand County, a landscape in which loss and gain are inextricable.

Although Vermont was the fastest-growing state right before the War of 1812, two of every five Vermonters departed in the period between 1850 and 1900. The untilled slopes that were such a draw to immigrants after the Revolution often turned out to be so flinty that the hill farmers' yields could not repay their labor. With the opening of Iowa Territory, many Vermont farmers decided that they would rather sail their plows across a sea of topsoil than break them on a boulder shoal. I read in Charles Morrissey's history of the state that 75% of Vermont had been cleared by 1850, as dairy farming and sheep raising prevailed. But unprofitable farms soon began to be abandoned and forests reclaimed the mountains. Today, 75% of our state is once more woodland. During the decades in which many parts of the country were ravaged by the mandates of prosperity, Vermont grew wilder and greener every year.

Economic stagnation protected the unspoiled countryside, with its network of villages. But there was too much suffering in these failed homesteads to allow for easy celebration, and a legacy of poverty remains in many of the hill towns like Bristol. What's more, the pleasant stability of Vermont is being pressed hard today. The telecommunications revolution, with the decentralized way of doing business it makes possible, turns quiet little worlds like this into targets for settlement and exploitation more abrupt than anything Vermont saw in its heyday a century and a half ago. I expect the next subdivision between Bristol and Burlington to be called "Sand County Estates." This winter, as snow briefly muffles the sounds of construction along Routes 7 and 100, offers a chance to retell the story of Vermont, in preparation for the coming season of distraction.

The wilderness areas designated in Vermont during the past 14 years are the climax of a century of enhancement through impoverishment. The Eastern Wilderness Act of 1975 (inspired in large part by Vermont's George Aiken) set aside lands that, while not pristine or vast like wilderness in the West and Alaska, still possessed natural qualities worthy of preservation. The Bristol Cliffs Wilderness behind my house was both farmed and thoroughly cut over, and it still turns up the stone and metal testaments of previous owners. This is no "virgin wilderness." Looking at the USDA *Soil Survey of Addison County*, it's also not too hard to see why our little ridge of wilderness was abandoned in the first place, allowing it to grow wild for this moment. While the fat soil to the west is identified on the survey maps by codes such as *Cw* and *VgB*—"Covington and Panton silty clays" and "Vergennes clay, 2 to 6 percent slopes"—Bristol Cliffs is scarified by labels like *LxE*—"Lyman-Berkshire very rocky complex, 20 to 50 percent slopes"—or simply *Rk*.

Such bony land along the heights is from one view just a discarded scrap. But a map of the state shows that our six wilderness areas—a total of 58,000 acres distributed along the Green Mountains' north–south axis—are also Vermont's green heart. Bristol Cliffs, Bread Loaf, Big Branch, Peru Peak, Lye Brook, and the George D. Aiken Wilderness (east of Bennington) focus a landscape where nature and culture have circled toward balance in a surprising, retrograde progression. By statute, Wilderness areas cannot be logged or built in. No new roads will be added, while existing ones will be allowed to fade away. As politicians consider plans for developing industry and broadening the tax base, these covenanted acres acknowledge a connection between Vermont's natural beauty and its century outside the mainstream. Just as valuable to me is the fact that the dense forests along these tracts are often fairly new, and that they are littered with reminders of previous chapters. They show that wildness can grow out of, and transform, the clearings of society. We don't always have to travel to Glacier National Monument or the Gates of the Arctic to find wilderness; under certain circumstances, it can come home to where we live.

These are the ironies of wilderness in many parts of the Northeast. This region, which was among the first parts of the country to be heavily settled, is now growing wild. Failing enterprises clear the ground for a new attempt at balance with the natural environment. And the abundant rainfall here allows the landscape to reassert its own agenda with a quickness unimaginable in states west of the hundredth meridian.

During this past December's hike into Bristol Cliffs, when I finally found the place, I kept thinking about Ron LaRose's word "heighth," in his description of where to search for the wreck. It reminded me of some lines in Frost's poem "Directive": "The height of the adventure is the height / Of country where two village cultures faded / Into each other. Both of them are lost." New York's shut-down mine, the failed homesteads of Vermont, and the wreckage of the Helldiver all show where lives, and whole communities, have been lost. But they also point to places where human vestiges and the region's nonhuman life have begun to fade together, lost in an emerging balance of wilderness and culture.

For the past couple of years I have been playing weekly games of *go* with my friend Pete Schumer. Invented in China and refined in Japan, *go* is a board game in which opponents alternate placing round, flattened stones on 361 intersections. White and black stones swirl around each other as the players contend for dominance over territory in various portions of the board. A pattern emerges in which the stones of both colors combine—a beautiful, intricate design beyond competition or invention.

Aji is a concept in *go* that helps me understand the swirl of nature and wildness in Vermont. Sometimes a player turns away from an area of the board where the opponent's position has become dominant. But the seemingly abandoned stones retain *aji* within the opposing color's sphere of influence. This word, which comes from the Japanese term for a lingering "taste," describes the fact that a minimal presence can suddenly "come alive." Scattered, discounted stones are empowered, combining with new ones of the same color when the action spills back into that sphere after an engagement elsewhere.

The little patches of wilderness in our state have functioned as a kind of *aji*. Wildness spreads back down into the Champlain Valley from rugged heights like Bristol Cliffs—which were the first places abandoned by the farmers and loggers. The forests have been waiting to return, and with them, animals we had almost forgotten. Moose, whose population dwindled to a small group in the northeast corner of Vermont, are now appearing more frequently in Addison County, here at the western portion of mid-state. Harold Hitchcock, a retired professor of biology at Middlebury College, feels that even the panthers, as mountain lions are called in this region, have begun to reappear. He has made a hobby of interviewing everyone who claims to have seen a panther. Some of them, he feels, have only seen deer in the twilit woods. But others have described details that make him believe in what they say. The mountains come alive for me in a new way now—I know that I will probably never spot one of these keen-eyed shadows for myself, but one of them may at least sometime watch me.

Go, with its circularity and suspension, reflects my vision of wilderness and culture in New England. What began as an opposition has slowly turned into a balance. When I was in high school, living in California and paying my dues to Friends of the Earth, I played a lot of chess. That's how the wilderness movement felt then, too. Double ranks of chessmen squared off for an apocalyptic encounter. Black and white lines clashed at the center of the board, and the number of pieces diminished until one side obtained the leverage to force checkmate. But the number of stones in *go* increases steadily in a game. When all portions of a board have been tested and claimed or relinquished, the players decide that the game is at an end; there is no checkmate and, if the proper number of handicap stones were placed down at the beginning, the game is often within a couple of points of being a tie.

One reason why this Japanese metaphor for New England wilderness appeals to me is that both regions would be called the East in California. Thoreau pointed out this paradox of directions on a round planet when he wrote in *Walden* that he would only join the migration west if he could continue far enough to arrive at the East. This goes along for me with another ancient symbol originating in China: the emblem of yin–yang, in which the swimming curves of dark and light bend around to form one circle. At the heart of the circle's darker side is a white dot, while a seed of blackness nestles in the light. *Aji*, an intricate pattern of complementarity and balance.

Just as Bristol Cliffs has been a saving remnant in Vermont, a little patch from which the heights could once more grow toward wilderness, the plane on South Mountain is a different kind of redemptive trace. This wilderness will never be pristine again, anymore than the shattered Helldiver will ever fly. The machine guns will not be dredged up from the muddy margins of the pond. But a dialogue between wilderness and culture is what we need now anyway, not a resolution. It may keep us from drawing our boundaries too straight, and remind us that sometimes we must go down before we find our second chance. The western-based environmental movement has often asserted the value of "virgin wilderness." But Vermont's return to wildness around the wreck offers, instead, the image of a marriage. Not a dichotomy of male and female, public and private, but an ongoing process of accommodation among the many necessities of growth.

When I first hiked up into Bristol Cliffs last August, my image of the fallen plane screened out the world that I passed by. This morning, though, when what I see is the luminous green-on-black of my computer's screen, and when what I hear is the steady current of its fan, I recall the world I waded through, looking for something else.

Leaves were thick and dark overhead, making the forest a cool, flickering place. The warmth of decay had coaxed out mushrooms everywhere—dimpled and white, day-glo orange, spotted tan ones scattered around the stumps like toads. Stands of goldenrod and black-eyed susans grew in the clearings. But most of the woodland wildflowers were already gone by. In the dim light of the woods, only the whorled wood aster was blooming, and that in profusion. Vigorous whirls of foliage lifted up limp white petals that swirled, in their turn, round the loose-packed yellow of the central disk-elegant late bloomers, lingering among the berries.

And berries there were: baneberries like little clusters of white pop beads, trilliums with their single, maraschino-bright fruit, clintonia with their pairs and triplets of purple berries turning black. Wildflowers lingered in their progeny, the consummation of their moving on. The August woods retained a memory of July. And now, as the earth undertakes its cold passage through December, orbiting back toward June, I return in writing to the scene of my reiterated hike. Memory compounds and thickens like the second-growth woods above my Bristol home.

14

The Future of the Northern Forest: Putting All the Pieces Together

Stephen C. Trombulak and Christopher McGrory Klyza

There are clearly many different visions for the future of the Northern Forest. The perspectives offered in the preceding chapters give much insight into why the current debate over this region has been so contentious. Opinions and interpretations differ on the root causes of the region's current problems, the actual identity of the critical problems, and the best strategies to ensure a sound future for the region's people and forests.

Yet we believe that all of these disparate views, taken together, make a positive contribution toward the development of effective solutions to meet the needs of all—current societies, forest ecosystems, and future generations. The wide range of voices presented in this book gives the reader a unique opportunity to evaluate the assumptions, arguments, and conclusions that have played the major roles in shaping the debate over the Northern Forest so far, and that will likely shape the debate for many years to come. No one person is likely to articulate or advocate all of the elements that must be part of the optimal future. Unfortunately, the vast majority of the public usually only has access to sources of information developed by individual factions in this debate, which clearly represent only a narrow slice of the total range of views, particularly newsletters and magazines produced by environmental groups, the forest products industry, or property rights advocates. Furthermore, few members of the public have the ability to participate in the lengthy meetings and hearings called to discuss these issues. As a result, most people derive their information and opinions from media sound bites—short bursts of news chosen for maximum impact but stripped of the content, and context, necessary to understand their validity or value.

After reading these chapters, we believe that the reader is now better able to assess the claims and suggestions of the major perspectives and is in a better position to play a knowledgeable role in determining the strategies that will shape the future of this region. By focusing on the common ground that already exists among these views, identifying the areas of remaining disagreement, and challenging the validity of claims made by representatives of any one perspective, the reader can now, directly and through representatives in local government, Citizen Advisory Committees, and in the U.S. Congress, participate in this debate in a more constructive way than before.

What Have We Learned?

The preceding chapters have analyzed the Northern Forest—both past and present—in several critical ways, particularly ecological, cultural, economic, political, and ethical. Several authors point out the critical role ecological processes and the current environmental conditions ought to play in this debate. Steve Trombulak describes the basic processes that have operated over time and space to shape the ecological components of the region. The Northern Forest is not a single ecosystem, but is rather a politically defined region made of many different ecosystem types, particularly spruce–fir and northern hardwood forests. These ecosystems vary throughout the region, being different in Maine than in New York, and at high than at low elevations. This spatial variation makes it unlikely that single solutions will be appropriate over the entire region. They also have varied considerably over time. These forests have developed over several millennia following the retreat of the continental glaciers about 10,000 years ago, constantly changing in the face of climate change, local disturbances, and successional process. In addition to the dominant themes of ubiquitous change and variation, he points out that the region has experienced near-continuous occupancy by humans—first Native American cultures and then settlers from Europe. From the perspective of the integrity of the forest, what is at issue is not the presence or absence of humans, but rather the scale and force of human influence relative to the ability of the forest to adjust and persist.

The speakers for the Abenaki Nation offer the complementary view that a sound future for this region can only be achieved if we all work to regenerate the forests that have been lost in recent times, and if we end the pollution that degrades the air, water, and soil. Their overall view on the "nature" of the region is that it is bigger than any one species or society. They advocate that, in the future, we must regain our respect for its immensity and stop tinkering with what we don't understand.

Both Emily Bateson and Jamie Sayen argue that environmental conditions within the Northern Forest region are currently quite poor, precisely because of the influence of the more than 300 years of intensive timber harvesting. Sayen, in particular, identifies the loss or degradation of several ecosystem characteristics, including old growth, soil conditions, water quality, and many species of plants and animals. As part of a long-term strategy for ecological restoration, both Sayen and Bateson describe one dominant approach within the conservation community for long-term protection of biological diversity: the establishment of a system of unmanaged land, buffered from large-scale disturbance by a landscape that includes managed lands and connections to other core reserves. Both of these authors also argue that simple-minded perspectives on conservation will not be successful in the long term: Protection of biological diversity is not the same as maximizing habitat diversity or species number throughout an area. The context of natural conditions and scales in space and time must be clearly understood.

Most of the authors in this book offer insights on the cultural aspects of the Northern Forest debate, arguing forcefully for the need to consider the people of the region when discussing strategies for the future. First, the Abenaki offer advice gained from their thousands of years of inhabitation in the region. They offer a model for living with the land and underscore the necessity of education in addition to action. This, they point out, is necessary because the key to living with the land is people choosing to do the right thing, not government forcing them to do it.

Identifying cultural elements as being important in this debate makes explicit the fact that, although humans are a part of nature, the human species plays a disproportionately large role in determining what elements must be part of ultimate solutions. Other authors in this volume, particularly Emily Bateson, Jonathan Wood, Jamie Sayen, and Hank Swan, note that the needs, rights, and responsibilities of humans must always be kept in mind. The definitions and priorities among these elements find their roots in the body of law that has evolved throughout the history of European settlement here, and as such are as true an expression of the regional culture as can be found in any work of art or literature.

Beyond this, however, these authors all point out the need to expressly define which society we are dealing with: global, regional, or local. Wood describes the interconnections that exist between societies outside of this region—implicit in the concept of a global economy—and how, for good or ill, they influence each other. Sayen contrasts landowners who live in the region with those who do not. These authors, perhaps knowingly, reinforce the concept of "regional tribes" in human society, each with special knowledge and relationships to the land in which they live. This gives great support to the conclusions made by the Abenaki, who offer that their tribe's centuries of regional occupancy give them experience that may help other tribes live here well.

The issues that originally brought the Northern Forest into the political light were all economic. From Tom Carr we learn that the forest products sector in the Northern Forest is very important to local economies, is important to regional economies, and is less significant at the national level. In Maine, which dominates the pulp and paper sector (62% of the output) and the lumber sector (50% of the output) in the four states, forest products are especially important. In the other three states, the forest products sector tends to be locally important. At the national level, the Northern Forest states supply less than 3% of the lumber output and not quite 9% of the pulp and paper output in the country.

In terms of both jobs and payroll in the four states, the forest products industry is of paramount importance compared to tourism and real estate/construction. It should be noted that although tourism employs half the number of people as the forest products industry, the tourism payroll is only approximately a fifth of that of the forest products industry. This underscores a potential problem with the push to a tourism-based economy. Finally, in the Northern Forest counties, 15% of the housing is vacation homes, 20% when we examine just Maine, New Hampshire, and Vermont. This suggests that second home development is not just coming; it is already here.

Both Jonathan Wood and Hank Swan offer perspectives on the role of timber in the region's economy—perspectives derived from their many years of direct experience. Wood perceives a great inequity between the benefits this industry brings to the region and the view held by some that the timber products industry needs to be reformed and regulated. He forcefully argues that the timber products industry not only plays a vital role here, but has done so in socially responsible ways with little compensation or understanding from the public. In addition to the direct contribution of the industry to the region's payroll, it also indirectly contributes to payroll (in the form of secondary and value-added jobs), recreational opportunities, protection of environmental values, and overall quality of life. Similarly, Swan advocates increased support for the forestry sector of the economy in the future, based on its demonstrated ability and potential to offer economic opportunities derived from local, renewable resources. These authors also offer their perceptions on the barriers to improved health of the industry. Both identify the need for general tax reform to remove the disincentives for keeping productive forest land as forests.

Swan goes on to suggest that past timber harvesting practices have greatly reduced the quality of the forests in many places in the Northern Forest, making forestry unprofitable there today and for the near-term future. This observation clearly points out the interconnectedness of all of these components, in this case the interplay between ecological and economic conditions.

Bateson and Sayen also contribute to this theme in their analysis of the region's economy. In their view, the social and economic force of the timber products industry, despite its potential to make major contributions to the people of this region, has so far been a negative influence, particularly through its emphasis on maximization of short-term profit at the expense of the workers, environmental integrity, and long-term sustainability. Sayen further describes how the predominance of transnational companies in this region results in a large flow of capital out of the area, as timber is converted into money that is then redirected to corporate operations in other parts of the country. He indicates the need to enhance local ownership, local employment, and value-added manufacturing to enhance the role of forestry in the region's economy.

The political perspective on this issue is also reviewed by several authors. In his chapter sketching the political context for the Northern Forest policy process, Chris McGrory Klyza examines the key problems, the political context, and the alternatives being discussed. The problems include the changing nature of the forest products industry (including mergers and acquisitions, increased lands for sale, and changing forest practices) as well as increased land for sale for development. It is these problems, and the perception of them, that have led to the rise of the Northern Forest on the policy agenda. Among the key characteristics of the political context have been the public sympathy toward environmental protection in New England, the rise of the property rights movement, the power of the forest products industry in Maine, and the economic boom and recession in New England in the 1980s and 1990s. The chief alternatives being considered involve land protection through easements, acquisition, or the use of green line reserves, and assorted changes in the tax systems to make forest ownership more attractive.

As a member of the Governors' Task Force and a longtime player in and analyst of the natural resources arena in New England, Carl Reidel's commentary on the political process thus far is rich in insights. Perhaps the greatest message we can take from his chapter is that agreement can be elusive when dealing with controversial issues. And when this is the case, task forces tend to agree on the least controversial recommendations rather than charting innovative, expansive policy directions. Reidel concludes that we are limited by time in dealing with the complexities of property rights and land values, and the federalist system makes it difficult to try innovative cooperative land management agreements. Both of these points should serve as political cautions as the process continues.

From John Collins we learn how the 6-million-acre Adirondack Park works. This mix of public and private land is the nation's largest working green line and offers one prospective model for land-use planning throughout the Northern Forest. In his analysis, the keys to the success of the park have been the "forever wild" protection of the public lands, the protection of the working forest on private lands, and the need for mixed economies in the park. The most pressing change to improve the working of the park is a simplification of government and government regulations, but not necessarily deregulation.

A combination of overdevelopment and decline is the reality facing local communities in the Northern Forest, according to Brendan Whittaker. We must be aware of both of these problems as we seek to craft a viable future for the Northern Forest. Since the towns will feel the effects of any new policies first, and since they will fight strongly for some degree of local control in the process, we must be sensitive to their needs and concerns. For example, we need to know what effect potential increases in public ownership will have on town revenues received from property taxes before pursuing this option, and how local communities can be included in planning for public purchase of lands.

Finally, we learn of the ethical dimensions to understanding this issue. From the Abenaki, we learn the importance of examining change in a long-term context, as they do, and relatedly, the value of patience. Their point that no one owns the land is also one that can lead us to examine the Northern Forest from a different perspective. It is also clear that the Abenaki need to be included in any conversation about the future of the Northern Forest, something that has not been apparent in the policy discussions thus far.

Stephanie Kaza presents an overview of five ethical polarities that have characterized the debate over the Northern Forest. She concludes by presenting a series of guidelines to help us move away from conflict and toward an incorporation of ethical considerations in the policy process. Perhaps most importantly, she reminds us of the importance of viewing power relations from an ethical perspective. The resiliency of nature, a sense of humility in the face of nature, and the variety of time horizons important in the interaction of wildness and culture are crucial lessons in John Elder's chapter.

What Are the Points of Consensus?

Out of this diversity come several points of agreement. We believe that these points provide a framework for long-term strategies for the future. Two things upon which all of these authors agree are the importance of maintaining (1) the ecological health of the forests and (2) the health of the forest products industry. These are linked in that, given the most likely scenarios for the future of this region, we cannot achieve one without the other. Of course, some of the factions in this debate still differ in their interpretation of what must be done to achieve these goals. Still, we believe that it is significant that in all of the voices heard in this book, not once was it suggested that either of these goals was not vital. It would greatly improve the tone of the public debate if all of the participants more openly expressed their support of these goals, committed themselves to supporting whatever is necessary to achieve them (even if it runs counter to personal biases and preferences), and in good faith worked toward implementation strategies. If this occurred, the debate could then focus more on what is needed to achieve mutually agreed upon goals and less on not giving up concessions to opposing factions.

The third point of consensus is clearly the support expressed by all for the importance of meeting the needs of the local residents of the region. No one has suggested that humans be removed from the area, that people be allowed to go hungry and homeless, or that their quality of life be ignored. Implicit in this point is the concept that the needs of the local residents outweigh the needs of the people outside of the region, be they timber workers in neighboring countries, corporate stockholders in other states, or consumers of timber products around the world. This is not to deny that the people who live here have responsibilities to others elsewhere, but no one should expect that these responsibilities will be met at the expense of basic dignity and cultural traditions.

The challenge for all of us now lies in crafting specific alternatives that build on and strengthen these three points of general agreement. To achieve this, all participants in the general dialog need to commit themselves to setting aside personal biases and expectations, and developing strategies that work for all.

Differences of Opinion

Despite the broad consensus expressed by these authors, however, there are still several points of disagreement that stand in the way of forging mutually satisfactory alternatives. Without careful assessment of these points, the reader runs the risk of concluding that the whole debate is intractable and no conclusions can be drawn. We urge the reader to look carefully at these differences of opinion and interpretation and, rather than ignore them, assess which position is best supported by data and observation.

Several of these differences stand out as major barriers. First, some of the authors disagree on the actual ecological health of the forest ecosystems in the Northern Forest, and to what extent timber-harvesting practices have degraded environmental conditions here. Jonathan Wood, for example, feels that environmental conditions are, in general, excellent and that this is a testament to the sound stewardship of many generations of people involved in timber harvesting. Jamie Sayen, on the other hand, argues that the ecological integrity of the Northern Forest has never been worse, due in large part to the abusive practices of the timber products industry. Can they both be right? Is it possible to find solutions acceptable to both the timber products industry and the environmental community if this point is not resolved?

We feel the answer to both of these questions is no. Everyone involved in this debate must assess the evidence presented to support each of these claims, and then stand firmly to promote that view in further debate. Assessing these claims, however, is not easy, and requires careful review of the data presented as evidence, as well as review of the data that is *not* presented. We are not completely unbiased in this debate ourselves, and it would not be appropriate for us to suggest what the reader should conclude. Rather, we wish to point out some of the elements that the reader should consider in forming a conclusion.

Sayen bases his contention on several factors, including the declining size of trees since the onset of timber harvesting (for example, a decrease in size of the average sawlog from 343 board feet in 1833 to 58 board feet in 1915), the near-complete absence of old growth, anecdotal comparisons of soil conditions following timber harvesting, degradation of water quality in many of the region's major rivers, and the decline and loss of several species and ecosystems throughout the area. Although he documents all of these changes, it is not clear from his analysis whether these have regional or long-term impacts (e.g., how quickly do soils recover following clearcutting?) or whether these changes were actually due to timber harvesting (e.g., all locally extirpated game animals were removed by hunting, and not clearcutting; similarly, most threatened ecosystems are not forests, but rather are rare ecosystems, such as alpine plant communities, that have declined due to climate changes and overuse by tourists).

Wood, on the other hand, draws his conclusions based on anecdotal observations that some species have increased their abundances in this region in recent times, and that the natural regenerative ability of these forests, exemplified by the recovery of the forests in Vermont following the extensive harvesting of the last century, surely demonstrates that little can permanently harm them. He argues that the extensive greenery and growing populations of many species on the hillsides of Vermont are an indication of growing, healthy forests.

Resolution of this disagreement will come when the participants in the debate weigh these competing measures of forest health, and begin to assess the condition of ecosystem characteristics for which neither author presents direct evidence: recovery rates of ecosystems following timber operations, changes in genetic diversity (a component of species resistance to disease and disturbance), maintenance of ecosystem processes—such as succession and soil production—under different management schemes, and the actual effect of managed landscapes on overall biological diversity.

Similarly, strong disagreement exists on (a) the responsibilities of private landowners to enact solutions, (b) how large a factor public ownership of land should be (such as expansion of the National Forests), (c) how compliance with conservation goals should be achieved, (d) whether restructuring property tax assessments on forest land is important independent of the implementation of other solutions or whether it should be promoted only on quid pro quo basis, and (e) whether regulations and/or incentive programs based on geographic areas—such as the green line approach described by John Collins—are advisable. How much of the opinions voiced here is based on fact, and how much is based on personal bias? What is the evidence for suggested consequences of any action? As one example, what is the evidence for community and ecological disruption if the Green and White Mountain National Forests are expanded to include larger portions of the Northern Forest region? Does the evidence clearly support the idea that this land would then stop supporting a timber-harvesting economy or ecologically healthy forests? Or is this statement rather an opinion based on a bias against public ownership? The reader, as well as everyone involved in the debate, must decide.

A Framework for Action

We find much merit in the arguments presented by all of the authors, and hold the view that forward movement is close at hand if a few critical barriers can be overcome. We point these out not to find fault with any one view or blame any particular faction for lack of progress on the issue, but rather to suggest areas that need to be resolved quickly if the opportunity presented by the current, but possibly brief, attention focused on this region is not to be squandered away. These points come in no particular order, as all are essential to building upon the consensus we already have.

1. Do not artificially confine the agenda of the debate. It has been apparent, both in the political history of the Northern Forest Lands Study and Council and in the views presented in the chapters of this book, that some people would like to direct the exploration for solutions away from particular topics by ruling them out-of-bounds, and enforce their will through use of the "minority rule" feature of the consensus method. Several examples are obvious, such as discussion of a green line approach for the Northern Forest, review of timber harvesting practices, the ecological dimension of the issue, and federal acquisition of land. The development of solutions to social problems is not helped when issues are ignored or suppressed from the outset. If they have no merit, then free and open discussion will quickly demonstrate that. If they have merit but are not useful or desirable in the specific context, then the fastest way to dispense with them will be to debate their merits and then move on. To suppress topics because they are undesirable to one or a few factions is not in the interests of society at large. Rather, it is merely self-serving.

Relatedly, it is clear that in many ways the Northern Forest itself is artificially defined. Throughout this book, most of the authors have used the boundaries established by the Forest Service and the Governors' Task Force. These boundaries, however, were drawn based on land ownership and political considerations. A more complete discussion of the Northern Forest would include lands in Canada, much more of New Hampshire, New York, and Vermont, and parts of western Massachusetts. Given existing political reality and the existence of the NFLC, though, we felt that it was best to focus on the existing boundaries, artificial as they, or any boundaries, may be.

2. Be clear on what you are asking for. Most of the rhetoric from some of the factions has focused on expression of dissatisfaction with other points of view without critical assessment of their own. This can be demonstrated by simplifying, perhaps unfairly, some of these arguments. For example, the environmental community feels that the timber products industry has created an unhealthy forest, without themselves identifying what a healthy forest would look like. The forest products industry, on the other hand, feels that the forests are perfectly fine, without ever indicating how they would recognize a loss of environmental integrity, or still further, how they would ever determine that they were responsible for damage and that they were not being careful stewards of the land. The claims of each are difficult to accept at face value if they leave the impression that any faction would never recognize when a problem really does or does not exist. Similarly, property rights advocates regularly point out the rights granted in the Constitution without also recognizing and responding to the limitations that go with those rights. These limitations have been repeatedly articulated by the U.S. Supreme Court, such as the government's power to enact zoning regulations, restrict development rights, and reduce an individual's profit earned from land speculation.

3. *Be consistent in the basis for your arguments.* It does not help the search for viable alternatives when a faction changes its opinions based on what serves its needs at the time. Should the federal government have a role in the Northern Forest? The answer cannot be yes when we talk about tax-supported benefits like road building but no when we talk about tax-supported benefits like acquisition of public land. Should landowners be responsible for the ecological integrity of their land? The answer cannot be yes when we talk about forest landowners and no when we talk about urban homeowners. Do we have social responsibilities that outweigh personal rights? The answer cannot be yes when we talk about the importance of Northern Forest timber to a global society but no when we talk about the actions of individual property owners. Should the rights of private property owners be respected? The answer cannot be yes when we talk about the descendants of European colonists and no when we talk about the descendants of those who were already here when the colonists arrived. Too often in this debate, "fundamental" beliefs are bent to the needs of the moment.

4. *Do not vilify opposing viewpoints through unrealistic characterizations.* Is it really true that the timber products industry is all represented by hard-working local Yankees and the environmental community is all represented by urbanites from outside of the region? Are there no environmentalists that live here? Are there no timber companies that make decisions in this region based on their national, rather than local, needs? Is it really true that transnational companies care less for local communities than do locally owned businesses because they can always uproot their operations and exploit a new region? Insider/outsider delineations are not productive, and are almost always wrong. This is evidenced by the facts that many of the vocal environmentalists engaged in this debate are residents of the Northern Forest region, and most of the transnational timber companies in this region are headquartered elsewhere.

5. *Do not avoid potential actions because of a fear of the "slippery slope."* We believe that future progress on the issues that face the Northern Forest will come from a mixture of alternatives. These alternatives

will not be developed, however, if potential elements are ruled out because they are related to strategies that are unacceptable. For example, is it wise to rule out a role for federal fee acquisition of land from willing sellers in supportive towns simply because federal taking of land through right of eminent domain or condemnation is undesirable? Similarly, is it wise to rule out public support of private business simply because of past abuses through the lack of adequate oversight and control? Slippery slopes that go from sound to abusive social systems will always exist, but they should not deter us from exploring the possibilities of such alternatives and developing strategies for avoiding unwanted consequences.

6. *Reject violence.* As pointed out by some authors, acts of violence have been committed by people who are trying to enforce their opinions and beliefs. We are a nation of laws. To advance a personal viewpoint by threatening or carrying out violence meant to intimidate or injure our fellow citizens is antithetical to every belief we hold as a people. (This contrasts sharply with our views on the role of intimidation and violence in our relations with other countries.) Those who conduct such practices must stop. Those who participate in this debate should view such practices harshly, and apply pressure to return the dialog to more ethically sound bases.

What then ought the future of the Northern Forest be? Given the broad consensus on the importance of ecological integrity, a forest-based economy, and meeting the needs of the local residents, can we derive a unified vision that works in the real world? Are the points of agreement enough to overcome the hostility and distrust that have grown out of this issue? Are the points of disagreement resolvable, or are they so major and the factions so set in their opinions that no negotiation is possible? Is it possible to meet the needs of the local people through a forest-based economy that does not harm the integrity of forest ecosystems?

We don't know the ultimate answers to these questions. Our goal—sustainable natural and human communities—is a difficult and far-reaching one. We are not quite sure what it looks like. But, we think that we must begin to move toward it, together, and that as we do so its contours will become clearer. In our opinion, it is vital that we try. To this end, we suggest that all participants in the debate over the future of the Northern Forest agree that a general structure for progress will strongly support the three areas of consensus, and to immediately develop specific strategies to achieve progress in small but clearly definable parts of the issue. Perhaps, for example, the ultimate solution will be a 20-point strategy for sustainability and cultural revitalization in the region, but we should not wait to debate the definitions of sustainability, what all 20 points are, and how all of them might relate to our cultural traditions before we implement any aspect of it. We must achieve success in a subset of the critical areas, such as tax restructuring and land acquisition, if we, as a society, are to feel empowered to continue with the debate in more challenging areas.

Finally, we must recognize that developing no strategies for the future, though an option, is not a truly viable approach. There are those who will argue for a maintenance of the status quo, who might stand to benefit from gridlock. These are people who might gain economically from the status quo (e.g., some developers, some forest products companies) or who are people resistant to change. But change has already come. More change is on the way. The increase in economic interconnections among regions around the globe, increasing human population size, increasing environmental deterioration, and the ever-accelerating evolution of American society guarantee that the people that live in the Northern Forest region cannot ignore the world and hope that the issues raised by the Northern Forest Lands Study, Governors' Task Force, and Northern Forest Lands Council will go away. We can either shape the future for ourselves—adapting our society and its traditions to ecological reality and global forces in positive ways—or we can wait passively for the future to be shaped for us. Gridlock on this issue will not result in the maintenance of the status quo. It will instead lead to increased development, unemployment, outside control of local economies, and forest degradation. The process of changing our views and working for the common need in good faith is often not pleasant, but it is better than the alternative.

Notes

1. A Natural History of the Northern Forest

1. This date is the estimate for when the continental ice sheets had left northern Maine, although glaciers were still present at higher elevations throughout the Northern Forest region. Jacobson, George L., Jr., and Ronald Davis, 1988, "Temporary and Transitional: The Real Forest Primeval," *Habitat: Journal of the Maine Audubon Society* (reprinted in *The Northern Forest Forum*, Spring Equinox, 1993, pp. 4–5).

2. Omernik, James M., 1987, "Ecoregions of the Conterminous United States," *Annals of the Association of American Geographers*, 77:118–125. The map *Ecoregions of the United States* (1986), also by Omernik, appears as a supplement to the 1987 article.

3. The regionalization scheme developed by the Office of Migratory Bird Management of the Fish and Wildlife Service has not been separately described as the topic of its own publication. However, the scheme is used to interpret data from the North American Breeding Bird Survey and can be found in any of the survey's annual summaries (e.g., Droege, Sam, and John R. Sauer, 1989, "North American Breeding Bird Survey Annual Summary 1988," *U.S. Fish and Wildlife Service, Biological Report*, 89 [13]).

4. Harper, Stephen C., Laura L. Falk, and Edward W. Rankin, 1990, *The Northern Forest Lands Study of New England and New York*, Rutland, Vt.: U.S. Department of Agriculture, Forest Service.

5. Bailey, Robert G., 1976, "Ecoregions of the United States (map)," Ogden, Utah: U.S. Department of Agriculture, Forest Service.

6. Johnson, Charles W., 1980, *The Nature of Vermont: Introduction and Guide to a New England Environment*, Hanover, N.H.: University Press of New England, pp. 39–40.

7. Cronon, William, 1983, *Changes in the Land: Indians, Colonists, and the Ecology of New England*, New York, New York: Hill and Wang, pp. 47–50.

8. Cronon, 1983, p. 42. This value is derived from Cronon's estimate of 41 persons per 100 square miles in Maine and applied over the entire Northern Forest region.

9. Cronon, 1983, p. 19.

10. Harper, Falk, and Rankin, 1990, p. 1.

11. Various aspects of the weather in the northeast are summarized in Lull, Howard W., 1968, *A Forest Atlas of the Northeast*, Upper Darby, Pa.: U.S. Department of Agriculture, Forest Service; and in Marchand, Peter, 1987, *North Woods: An Inside Look at the Nature of Forests in the Northeast*, Boston: Appalachian Mountain Club Books.

12. Summarized from the Soil Conservation Service by Lull, 1968, pp. 14–15.

13. These general descriptions are drawn from several sources, particularly *Northeastern Area Forest Health Report*, 1992, NA-TP-03-93, Radnor, Pa.: U.S. Department of Agriculture, Forest Service; Johnson, 1980; and Marchand, 1987.

14. Cronon, 1983, pp. 109–110.

15. Jackson, Stephen T., and Donald R. Whitehead, 1991, "Holocene Vegetation Patterns in the Adirondack Mountains," *Ecology*, 72:641–653; Jacobson and Davis, 1988.

16. Lorimer, Craig, 1977, "The Presettlement Forest and Natural Disturbance Cycles of Northeastern Maine," *Ecology*, 58:139–148.

17. Marchand, 1987, pp. 96–98.

18. Harper, Falk, and Rankin, 1990, pp. 23–24.

19. Marchand, 1987, pp. 111–125.

20. I have calculated these figures from the following field guides and atlases: Burt, William H., and Richard P. Grossenheider, 1976, *A Field Guide to the Mammals*, Boston: Houghton Mifflin; DeSante, David, and Peter Pyle, 1986, *Distributional Checklist of North American Birds*, Lee Vining, Calif.: Artemesia Press; Conant, Roger, and Joseph T. Collins, 1991, *Reptiles and Amphibians: Eastern and Central North America*, Boston: Houghton Mifflin.

21. Mills, L. Scott, Michael E. Soule, and Daniel F. Doak, 1993, "The Keystone-Species Concept in Ecology and Conservation," *BioScience*, 43:219–224.

22. Harper, Falk, and Rankin, 1990, p. 25.

23. Cronon, 1983.

24. Johnson, 1980, p. 44.

25. Judd, Richard, 1989, *Aroostook: A Century of Logging in Northern Maine*, Orono, Maine: University of Maine Press.

26. Johnson, 1980, p. 50.

27. Cameron Duffy, David, and Albert J. Meier, 1992, "Do Appalachian Herbaceous Understories Ever Recover from Clearcutting?" *Conservation Biology*, 6:196–202.

28. United States Forest Service, 1985, "Results of Recent Research Related to Whole-Tree Harvesting for the Northeastern Forest Experiment Station," Orono, Maine: U.S. Department of Agriculture, Forest Service.

29. Lansky, Mitch, 1992, *Beyond the Beauty Strip: Saving What's Left of Our Forests*, Gardiner, ME: Tilbury House, pp. 170–174.

30. Lansky, 1992, p. 209.

31. Harris, Larry D., 1984, *The Fragmented Forest*, Chicago: University of Chicago Press. This entire volume reviews the ecological consequences of fragmentation on forest ecosystems that are left standing.

32. Wilcove, David S., Charles H. McLellan, and Andrew P. Dobson, 1986,

"Habitat Fragmentation in the Temperate Zone," in *Conservation Biology: the Science of Scarcity and Diversity*, edited by Michael E. Soule, Sunderland, Mass.: Sinaeur Associates, pp. 237–256.

33. Webb, William L., Donald F. Behrend, and Boonraung Saisorn, 1977, "Effect of Logging on Songbird Populations in a Northern Hardwood Forest," *Wildlife Monograph*, 55:1–35.

34. Terborgh, John, 1989, *Where Have All the Birds Gone?* Princeton, N.J.: Princeton University Press; Bohning-Gaese, Katrin, Mark L. Taper, and James H. Brown, 1993, "Are Declines in North American Insectivorous Songbirds Due to Causes on the Breeding Range?" *Conservation Biology*, 7:76–86.

35. See, for example, Johnson, 1980, for the history of extirpations of these species in Vermont.

36. Brock, Ranier, 1992, "Proceedings from the Biological Resources Diversity Forum," Concord, N.H.: Northern Forest Lands Council.

37. *Northeastern Area Forest Health Report*, 1992.

38. Carpenter, Stephen R., 1980, "The Decline of *Myriophyllum spicatum* in a Eutrophic Wisconsin Lake," *Canadian Journal of Botany*, 58:527–535.

3. The Northern Forest

1. Harper, Stephen C., Laura L. Falk, and Edward W. Rankin, 1990, *The Northern Forest Lands Study of New England and New York*, Rutland, Vt.: U.S. Department of Agriculture, Forest Service. The Adirondack State Park, 6 million acres of private and public land, has an interesting and complex history independent of the Northern Forest issue. For a history of the park, see Graham, Frank, 1978, *The Adirondack Park: A Political History*, Syracuse, N.Y.: Syracuse University Press.

2. Data in Table 1 from the *Bangor Daily News*, July 11–12, 1992, as cited in *The Northern Forest Forum*, Autumn 1992, p. 23; *The Northern Forest Advocate*, Winter 1992, p. 5.

3. Stegner, Page, 1991, "Let It Be Woods," *Sierra*, September/October, p. 60.

4. The analytic approach used in this chapter is based on Kingdon, John W., 1984, *Agendas, Alternatives, and Public Policies*, Boston: Little, Brown.

5. Sedjo, Roger A., 1991, "Forest Resources: Resilient and Serviceable," in *America's Renewable Resources: Historical Trends and Current Challenges*, edited by Kenneth D. Frederick and Roger A. Sedjo, Washington, D.C.: Resources for the Future, pp. 102, 110. This increase in timber prices could lead to the increased viability of the timber industry in the northeast. It could also lead to another wave of corporate takeovers, since the standing timber will be even more undervalued than it was in the 1980s.

6. Sedjo, 1991, pp. 100–102.

7. Sedjo, 1991, pp. 104–105.

8. Sedjo, 1991, p. 109. For a general overview of the New England forest, see Irland, Lloyd C., 1982, *Wildlands and Woodlots: The Story of New England's Forests*, Hanover, N.H.: University Press of New England.

9. Boucher, Norman, 1989, "Whose Woods These Are: The Task of Saving the Last of Northern New England's Wildlands Is a Puzzle of Possibilities, Contention, and Promise," *Wilderness*, Fall, pp. 20–21; Northern Forest Lands Council, 1992a, "Forum on Land Sales of Coburn Lands Trust and Former Diamond International Corporation," Concord, N.H., p. 2.

10. Boucher, 1989, pp. 21–23; Helm, Leslie, 1988, "In New England, the Axman

Cometh," *Business Week*, October 3, pp. 66–68; Northern Forest Lands Council, 1992a, pp. 2–3.

11. Boucher, 1989, pp. 21–24; Helm, 1988, pp. 66–68.

12. Boucher, 1989, p. 25; Northern Forest Lands Council, 1992a, pp. 3–5; Stegner, Page, 1990, "Fate of the Northeast Kingdom," *National Parks*, January/February, p. 28.

13. Commission on the Adirondacks in the Twenty-First Century, 1990, *The Adirondack Park in the Twenty First Century*, Albany: State of New York; Mitchell, John G., 1989, "A Wild Island of Hope: The Adirondack Park Faces Its Own Set of Problems," *Wilderness*, Fall, p. 50; Northern Forest Lands Council, 1992a, pp. 3–5.

14. "Section I. Fragments and Systems," 1989, *Wilderness*, Spring, p. 22; "Sierra Club Public Lands Campaigns: Northeast Forests," 1989, *Sierra*, September/October, p. 64.

15. Helm, 1988, pp. 66–68.

16. Stegner, 1990, p. 27.

17. "Bowater Buys Great Northern," 1991, *New York Times*, October 11, p. D4; Helm, 1988, pp. 66–68.

18. See, for example, Sierra Club, n.d., "Hostile Takeovers in the Forest Industry," Saratoga Springs, N.Y.: Sierra Club Northeast Office.

19. Wuerthner, George, 1988, "Northeast Kingdoms: Is It Time to Rescue the Last of New England's Wilderness?" *Wilderness*, Summer, p. 47; Boucher, 1989, p. 37; Sierra Club, n.d., "Biodiversity in the Northern Forest Lands Study Area," Saratoga Springs, N.Y.: Sierra Club Northeast Office; Sierra Club, n.d., "Current Forestry Practice," Saratoga Springs, N.Y.: Sierra Club Northeast Office. For a comprehensive critique of industrial forestry as practiced in Maine, see Lansky, Mitch, 1992, *Beyond the Beauty Strip: Saving What's Left of Our Forests*, Gardiner, Maine: Tilbury House.

20. Boucher, 1989, pp. 36–37. In addition to the individual corporate and large private landowners, the forest products industry is represented by a number of organizations in the Northern Forest issue. These include the American Forest Council, American Forest Resources Alliance, Empire State Forest Products Association, Maine Forest Products Council, New Hampshire Timberland Owners Association, and Vermont Forest Products Association.

21. Council on Environmental Quality, 1980, *Public Opinion on Environmental Issues*, Washington, DC: Government Printing Office, pp. 44–45; Echelberger, Herbert E., Albert E. Luloff, and Frederick E. Schmidt, 1991, "Northern Forest Lands: Resident Attitudes and Resource Use," Radnor, Pa.: U.S. Department of Agriculture, Forest Service, p. 11; "Maine Hoping to Preserve the Good Life with the Good Earth," 1987, *New York Times*, November 8, I, p. 59; Stegner, 1990, p. 27.

22. Of the four members of the NFLC from each state, one member should be appointed to represent each of the following constituencies: state government conservation agency (Maine Department of Conservation, New Hampshire Division of Forests and Lands, New York Department of Environmental Conservation, and Vermont Department of Forests, Parks and Recreation), forest products industry, environmental groups, and local communities.

23. Mitchell, John G., 1992, "Love and War in the Big Woods," *Wilderness*, Spring, p. 12.

24. See Klyza, Christopher McGrory, 1991, "Framing the Debate in Public Lands Politics," *Policy Studies Journal*, 19:577–585.

25. Allen, Susan, 1991, "Northern Forest Protection Issues Debated at Hear-

ing," *Rutland Herald*, July 16, p. 1; Austin, Phyllis, 1991, "Emotions Are High as Old Adversaries Confront the Northern Forest," *Maine Times*, August 30, pp. 4–6; Beamish, Richard, 1992, "Adirondack: The Centennial Wild," *Wilderness*, Spring, p. 23; Mitchell, 1992, p. 18; U.S. Congress, Senate, Committee on Agriculture, Nutrition, and Forestry, 1991, "Oversight on Forest Land Conservation and Related Economic Development within the Northern Forest Lands Study Area," 102d Congress, 1st Session, July 15. In addition to these local groups, national support has come from the National Inholders Association and the Center for the Defense of Free Enterprise.

26. Austin, Phyllis, 1992, "Mitchell, Cohen Withdraw Land Council Backing," *Maine Times*, March 6, p. 6; Sierra Club, 1991, "Northern Forest News," Saratoga Springs, NY, Summer. Despite the lack of congressional authorization, the NFLC has still been receiving federal funds to support its mission. It received $1.2 million for Fiscal Year 1992 and approximately the same for Fiscal Year 1993.

27. Commission on the Adirondacks in the Twenty-First Century, 1990; Beamish, 1992, p. 23.

28. Northern Forest Lands Council, 1992a, p. 3.

29. Wuerthner, 1988, pp. 45–51. For similar regionwide proposals, see "Section I. Fragments and Systems," 1989, p. 22; "Sierra Club Public Lands Campaigns: Northeast Forests," 1989, p. 64. In addition to these areas, the National Parks and Conservation Association favored a new park along the Machias River in Maine, Wild and Scenic River designation for the St. John River in Maine, and protection of the area around Cobscook Bay, Maine. See Stegner, 1990, p. 29. As discussed, the Nash Stream drainage in New Hampshire was purchased in 1988, but it was not added to the White Mountain National Forest.

30. Watkins, T. H., 1989, "The Maine Woods Reserve," *Wilderness*, Fall, p. 41.

31. Mitchell, 1989, pp. 44–50.

32. Audubon, National Wildlife Foundation, Sierra Club, and Wilderness Society, 1991, "Saving the Northern Forest: An Issue of National Importance."

33. The groups in the Alliance at this time were the Adirondack Council, American Forestry Association, Appalachian Mountain Club, Appalachian Trail Conference, Association for Protection of the Adirondacks, Audubon Society of New Hampshire, Conservation Law Foundation, Maine Audubon Society, National Audubon Society, National Wildlife Federation, New Hampshire Wildlife Federation, Sierra Club, Society for the Protection of New Hampshire Forests, Trust for Public Land, and Wilderness Society. By December 1992, 13 new groups had joined the Alliance: the Adirondack Mountain Club, Environmental Air Force, Green Mountain Chapter of the Society of American Foresters, Green Mountain Club, Natural Resources Council of Maine, Natural Resources Defense Council, Nature Conservancy, New England Society of American Foresters, Preserve Appalachian Wilderness, Restore: The North Woods, Student Environmental Action Coalition, Vermont Land Trust, and Vermont Natural Resources Council.

34. See Elliott, Jeff, and Jamie Sayen, 1990, "The Ecological Restoration of the Northern Appalachians: An Evolutionary Perspective," North Stratford, N.H.: Preserve Appalachian Wilderness.

35. Harper, Falk, and Rankin, 1990, pp. 45–59; Governors' Task Force on Northern Forest Lands, 1990, "The Northern Forest Lands: A Strategy for their Future," Montpelier, Vt.: Vermont Department of Forests and Parks, pp. 5–11.

36. Harper, Falk, and Rankin, 1990, pp. 59–69; Governors' Task Force on Northern Forest Lands, 1990, pp. 5–7.

37. Northern Forest Lands Council, 1992b, "Northern Forest Update," 2:2.

The other two subcommittees of the NFLC are Biological Resources and Conservation Strategies. The NFLC has also published two reports not economically oriented: "A Summary of State Land Conservation Activities for the States of Maine, New Hampshire, New York and Vermont" and "Forum on Biological Resource Diversity." Interestingly, no scientists are on the NFLC or its staff; the alternatives being considered focus on economics, not biology.

38. See also Harper, Falk, and Rankin, 1990, pp. 40–45.

39. Harper, Falk, and Rankin, 1990, pp. 75–79.

4. The Northern Forest Economy

1. Harper, Stephen C., Laura L. Falk, and Edward W. Rankin, 1990, *The Northern Forest Lands Study of New England and New York*, Rutland, Vt.: U.S. Department of Agriculture, Forest Service.

2. Harper, Falk, and Rankin, 1990, pp. 4–5.

3. Pearce, P. H., 1967, "The Optimum Forest Rotation," *Forestry Chronicle*, 43:178–195; Samuelson, Paul A., 1976, "Economics of Forestry in an Evolving Society," *Economic Inquiry*, 14:466–492.

4. Harper, Falk, and Rankin, 1990, p. 5.

5. Hartman, Richard, 1976, "The Harvesting Decision When a Standing Forest Has Economic Value," *Economic Inquiry*, 14:52–58.

6. Harper, Falk, and Rankin, 1990, p. 26.

7. To avoid making arbitrary distinctions about the content level of wood or fiber in a given product (e.g., should furniture be included but not boxboard?), the forest products industry is narrowly defined as the Standard Industrial Classification categories of lumber and wood products and paper and allied products.

8. Irland Group, 1992, *Economic Impact of Forest Resources in Northern New England and New York*, Northern Forest Alliance, p. 33; U.S. Department of Commerce, Bureau of Economic Analysis, 1992, *Business Statistics*, Washington, D.C.: Government Printing Office.

9. U.S. Department of Agriculture, 1987, *Census of Agriculture*, Washington, D.C.: Government Printing Office.

10. Delorme Mapping Company, 1988 and 1990, *The Maine Atlas*; *The New Hampshire Atlas*; *The Vermont Atlas*, Freeport, Maine.

11. Schneider, Paul, 1992, "The Adirondacks: The Remaking of a Wilderness," *Audubon*, May/June 1992, p. 55.

12. Irland Group, 1992, pp. 40–41.

13. Irland Group, 1992, p. 38.

14. Northern Forest Lands Council, 1990, *Northern Forest Update*, November/December 1990, p. 2.

15. Harper, Falk, and Rankin, 1990, p. 111.

16. U.S. Bureau of the Census, 1990b, *Statistical Abstract of the United States*, Washington, D.C.: Government Printing Office, p. 439.

17. U.S. Bureau of the Census, 1990a, *Census of Population and Housing*, Washington, D.C.: Government Printing Office.

18. Harper, Falk, and Rankin, 1990, p. 1.

19. World Resources Institute, 1992, *World Resources 1992–93*, New York: Oxford University Press, p. 288; World Resources Institute, 1990, *World Resources 1990–91*, New York: Oxford University Press, p. 292.

20. U.S. Bureau of the Census, 1990b, p. 672.

21. Value Line, 1992, "Paper and Forest Products Industry," *The Value Line Investment Survey*, New York.
22. Keefe, Sally, 1993, "Leverage and Productivity in the U.S. Forest Products Industry," senior thesis, Middlebury, Vt.: Middlebury College.
23. Sedjo, Roger, and Kenneth Lyon, 1990, *The Long-Term Adequacy of World Timber Supply*, Washington, D.C.: Resources for the Future.
24. Pigou, A. C., 1932, *The Economics of Welfare*, London: Macmillan.
25. Coase, Ronald, 1962, "The Problem of Social Cost," *Journal of Law and Economics*, 3:1–44.
26. Friedman, Milton, 1962, *Capitalism and Freedom*, Chicago: University of Chicago Press, p. 31.
27. Weisbrod, Burton, 1964, "Collective-Consumption Services of Individual-Consumption Goods," *Quarterly Journal of Economics*, 78:471–477.
28. Krutilla, John, 1967, "Conservation Reconsidered," *American Economic Review*, 57:777–786.
29. Fischel, William, 1990, "Introduction: Four Maxims for Research on Land-Use Controls," *Land Economics*, 66:229.

5. Ethical Tensions in the Northern Forest

1. For an introduction to these regional forests, see Norse, Elliot A., 1990, *Ancient Forests of the Pacific Northwest*, Washington, D.C.: Island Press; Gradwohl, Judith, and Russell Greenberg, 1988, *Saving the Tropical Forests*, Washington, D.C.: Island Press; and Pavlik, Bruce M., Pamela C. Muick, Sharon Johnson, and Marjorie Popper, 1991, *Oaks of California*, Los Olivos, Calif.: Cachuma Press.
2. Harrison, Robert Pogue, 1992, *Forests: The Shadow of Civilization*, Chicago: University of Chicago Press.
3. Cronon, William, 1983, *Changes in the Land: Indians, Colonists, and the Ecology of New England*, New York: Hill and Wang.
4. For a case study of Maine's forests, see Lansky, Mitch, 1992, *Beyond the Beauty Strip: Saving What's Left of Our Forests*, Gardiner, Maine: Tilbury House.
5. Discussed in "Adirondack Voices," the newsletter of the Residents' Committee to Protect the Adirondacks.
6. Partridge, Ernest, ed., 1981, *Responsibilities to Future Generations*, Buffalo, N.Y.: Prometheus Press.
7. Macy, Joanna, 1991, *World as Lover, World as Self*, Berkeley, Calif.: Parallax Press, pp. 206–291.
8. For a discussion of the universalizing tendency, see Cheney, Jim, 1989, "Postmodern Environmental Ethics: Ethics as Bioregional Narrative," *Environmental Ethics*, 11:117–134.
9. Narayan, Uma, 1988, "Working Together Across Differences: Some Considerations on Emotions and Political Practice," *Hypatia*, 3(2):31–47.
10. See "Whose Common Future?" 1992, *The Ecologist*, 22(4), entire issue on the enclosure of the commons.
11. Snyder, Gary, 1990, "The Place, the Region, and the Commons," in *The Practice of the Wild*, San Francisco: North Point Press, pp. 25–47.
12. "Whose Common Future?" 1992, p. 124 for a description of insider/outsider dynamics in Third World street markets.
13. Berry, Wendell, 1992, "Conservation is Good Work," *Wild Earth*, Spring, p. 82.

14. Leopold, Aldo, 1949, "The Land Ethic," in *A Sand County Almanac*, New York: Ballantine Books, p. 239.

15. For a new interpretation of corporate responsibility to private rights, see Saxe, Dianne, 1992, "The Fiduciary Duty of Corporate Directors to Protect the Environment for Future Generations," *Environmental Values*, 1:243–252.

16. Stone, Christopher, 1974, *Should Trees Have Standing? Toward Legal Rights for Natural Objects*, Los Altos, Calif.: William Kaufmann.

17. Leopold, 1949, p. 238.

18. Lansky, 1992, pp. 137–162 on industrial management choices, for example.

19. The American Forestry Association has been working on an ethical code for several years; several other codes for biologists and ecologists are in committee.

20. For examples of this healing work, see the journal of the Society for Ecological Restoration, Restoration and Management Notes (Madison: University of Wisconsin Press).

21. "Power: The Central Issue," 1992, *The Ecologist*, 22(4):157–164.

22. Lansky, 1992, p. 415.

6. The Political Process of the Northern Forest Lands Study

1. Portions of this chapter are based on Reidel, Carl, May/June 1990, "The Northern Forest: Our Last Best Chance," *American Forests*, 96(5–6):22–25, 75–77; Reidel, Carl, March/April 1993, "Endgame in the Northern Forest," *American Forests*, 99(3–4):29–32, 55–56; and a paper given at the 1992 University of Vermont Aiken Lecture Series.

2. Northern Forest Lands Council, March 13, 1992, *Summary of Proceedings: Forum on Land Sales of Coburn Lands Trust and Former Diamond International Corporation*, Concord, N.H.: Northern Forest Lands Council.

3. Northern Forest Lands Council, March 13, 1992, p. 2.

4. Hagenstein, Perry, August, 1987, *A Challenge for New England: Changes in Large Forest Land Holdings*, Boston: The Fund for New England and The New England Natural Resources Center.

5. Northern Forest Lands Council, March 13, 1992, p. 3.

6. Abbott, Will, November 12–13, 1992, "The Management of Nash Stream Forest: A Partnership Approach," *Proceedings of the George D. Aiken Lecture Series*, University of Vermont.

7. Governors Task Force Members: *Maine*: Edward Johnson, Executive Director, Maine Forest Products Council; C. Edwin Meadows, Commissioner, Department of Conservation; J. Mason Morfit, Executive Director, The Nature Conservancy. *New Hampshire*: Henry Swan, President, Wagner Woodlands; Will Abbott, Executive Director, Land Conservation Investment Program; Paul Bofinger, President, Society for the Protection of New Hampshire Forests. *New York*: Ross Whaley, President, College of Environmental Science and Forestry, State University of New York; Robert Bathrick, Director, Division of Lands and Forests, Department of Environmental Conservation; George Davis, Executive Director, Commission on the Adirondacks. *Vermont*: Peter B. Meyer, Vice President, E. B. Hyde Corporation; Mollie Beattie, Commissioner of Forests and Parks (later, Deputy Secretary), Agency of Natural Resources; Carl Reidel, Director, Environmental Program, University of Vermont.

8. Laura Falk and Ted Rankin were married in June 1989 in a ceremony that I conducted as a Vermont Justice of the Peace. Sadly, Ted died of cancer in April 1992. Laura now works on the White Mountain National Forest.

9. Harper, Stephen C., Laura L. Falk, and Edward W. Rankin, 1990, *The Northern Forest Lands Study of New England and New York*, Rutland, Vt.: U.S. Department of Agriculture, Forest Service, p. v.

10. Wuerthner, George, Summer, 1988, "Northeast Kingdoms," *Wilderness*, Summer 1988, pp. 45–51.

11. Most of the commissioned studies are summarized in appendices of Harper et al., 1990.

12. Binkley, Clark, and Perry Hagenstein, 1989, *Conserving the North Woods: Issues in Public and Private Ownership of Forested Lands in Northern New England and New York*, New Haven, Conn.: Yale School of Forestry and Environmental Studies.

13. Yaro, Robert D., July 1989, *The Northern Forest Lands: Greenline and Regional Planning Alternatives*, Amherst, Mass.: University of Massachusetts, Center for Rural Massachusetts.

14. Preserve Appalachian Wilderness (undated newsletter), c. December 1988, *The Working Forest Is Not Working: A Critique of the Northern Forest Lands Study*, North Stratford, N.H.

15. The Governors' Task Force on Northern Forest Lands, May 1990, *The Northern Forests: A Strategy for their Future—An Action Plan of The Governors' Task Force on Northern Forest Lands*, Conservation Education Section of the Vermont Department of Forests and Parks.

16. A joint policy statement of the National Audubon Society, National Wildlife Federation, Sierra Club, and the Wilderness Society (unpublished report), October 22, 1991, *Saving the Northern Forest: An Issue of National Importance.*

17. NFLC Members and staff: Executive Director, Charles A. Levesque; *State Coordinators*: Donald Mansius, Maine; Susan Francher, New Hampshire; Karyn Richards, New York; James Horton and Charles Johnson,* Vermont. *NFLC Members*: Maine: Jerry Bley, Natural Resource Consultant; Edward Johnson,* Maine Forest Products Council; Janice McAllister, Abbot Selectwoman; C. Edwin Meadows,* Maine Department of Conservation. New Hampshire: Paul Bofinger,* Society for the Protection of New Hampshire Forests; John Harrigan, landowner; Beaton Marsh, local representative; John Sargent, New Hampshire Division of Forests and Lands. New York: Robert Bendick, New York State Department of Environmental Conservation (Chair of NFLC); Robert Stegemann, International Paper; Barbara Sweet, Newcomb Councilwoman; Neil Woodworth, Adirondack Mountain Club. Vermont: Richard Carbonetti, Consulting Forester; Peter B. Meyer,* E. B. Hyde Corporation; Conrad Motyka, Vermont Department of Forests, Parks, and Recreation; Brendan Whittaker, Vermont Natural Resources Council. *U.S. Forest Service*: Michael T. Rains, Associate Deputy Chief, State and Private Forestry. (Asterisk indicates those involved in original Governors' Task Force.)

18. Northern Forest Lands Council, November/December 1992, "States Host Public Input Sessions," *Northern Forest UPDATE*, 2(5), p. 1.

19. From a statement in a brochure distributed by the NFLC. c. 1992. See also "Northern Forest Lands Council Mission Statement and Operating Principles."

20. Northern Forest Lands Council, November/December 1992, "Draft Interim Status Report Released, Council Solicits Public Input," *Northern Forest UPDATE*, 2(5), p. 1. See also May/June 1992, "Proposed FY 93 Activities," *Northern Forest UPDATE*, 2(2), p. 3.

21. The Northern Forest Alliance membership directory (December 1992) includes the Adirondack Council; American Forestry Association; Appalachian Mountain Club; Appalachian Trail Conference; Association for the Protection

of the Adirondacks; Audubon Society of New Hampshire; Conservation Law Foundation (of Boston); Environmental Air Force; Green Mountain Club; Maine Audubon Society; National Audubon Society; National Wildlife Federation; Natural Resources Council of Maine; New England Society of American Foresters; New Hampshire Wildlife Federation; Natural Resources Defense Council; Preserve Appalachian Wilderness (PAW); Restore: the North Woods; Sierra Club; Society for the Protection of New Hampshire Forests; Student Environmental Action Coalition; Green Mountain Chapter of the Society of American Foresters; The Nature Conservancy; The Wilderness Society; Trust for Public Land; Vermont Land Trust; Vermont Natural Resources Council.

22. Reidel, Carl, 1990, "Forest Policy and Land Use Planning; Trends in the United States," in *Land Use for Agriculture, Forestry, and Rural Development*, edited by M. C. Whitby and P. J. Dawson, Newcastle upon Tyne, England: The University, p. 272.

23. Reidel, Carl, and Jean Richardson, December 1992, "A Public/Private Cooperative Paradigm for Federal Land Management," *Multiple Use and Sustained Yield; Changing Philosophies for Federal Land Management*, Washington, D.C.: Prepared by Congressional Research Service, Library of Congress, for the Committee on Interior and Insular Affairs, U.S. House of Representatives, 102nd Congress, 2nd Session, Committee Print 11, pp. 163–164.

24. McClaughry, John, 1976, "Farmers, Freedom, and Feudalism: How to Avoid the Coming Serfdom," *South Dakota Law Review*, 21(3):156.

25. Reidel, Carl, May/June 1990, "The Northern Forest: Our Last Best Chance," *American Forests*, 96(5–6):22–25, 75–77.

8. A View from Local Government

1. I would like to thank Jerry A. Bley and Barbara Sweet, members of the Northern Forest Lands Council from Maine and New York, respectively, for their help in preparing this chapter. Jerry Bley is a member of the Conservation Commission and Comprehensive Planning Committee of his town of Readfield, Maine; Barbara Sweet is a Town Councilwoman from her Town of Newcomb, New York, and also serves as Chair of the Essex County (N.Y.) Industrial Development Authority.

2. The "Town" in New England usage includes not only the settled village areas but all of the acreage, including the most rural hinterlands. County governments, strong in most of the United States (including New York State), are quite vestigial in New England, perhaps having a court system, a sheriff's department, and, in New Hampshire, a nursing home or hospital. Most northern New Englanders identify their place, first by the state (e.g., "I'm a Vermonter", or "A State-of-Mainer"), then their town (e.g., "I'm a New Hampshirite, from up Errol way"). Only rarely will they mention county. New York State, including the Tug Hill–Adirondack Northern Forest regions, utilizes the county level more extensively (e.g., for highways).

3. See, for example, the Maine Natural Resources Council's proposal for a 7 to 9 million acre "North Woods Conservation Area" that would encompass almost 90% of Maine's 10.5 million acres in unorganized areas (*Maine Environment*: Bulletin of the MNRC, October/November 1992).

4. Personal telephone conversations with the Offices of the Vermont and Maine Secretaries of State, December 1992.

5. Johnson, Fenton, 1992, "In the Fields of King Coal," *The New York Times Sunday Magazine*, November 22, p. 30.
6. *Overview*, Northern Forest Lands Council.
7. *The Berlin (NH) Daily Reporter*, January 20, 1993, p. 1.
8. "Selectmen's Powers and Duties: An Overview," *Proceeding from a Conference at the Lake Morey (Vermont) Inn*, May 12, 1992, p. 4. "The courts have also recognized a higher standard of immunity for school directors than selectmen, based on the realization that school directors *are actually state officials* (emphasis added) since they are acting to fulfill the constitutional mandate that the state maintain a competent [*sic*] number of schools in each town."
9. Personal conversation with Ross Whaley, January 1993. Whaley also shared his concept of "the New England Town Meeting" model in his presentation at the November 1992 Aiken Lecture Series on the Northern Forest.
10. Keefe, Sally, 1992, *Northern Forest Project Newsletter (Vermont Natural Resources Council*, 1(1), September.

9. Sustaining Our Forest—Crafting Our Future

1. The World Commission on Environment and Development, 1987, *Our Common Future*, New York: Oxford University Press, p. 8.
2. Meadows, Donella, Dennis Meadows, and Jorgenm Randers, 1992, *Beyond the Limits: Confronting Global Collapse, Envisioning a Sustainable Future*, Post Mills, Vt.: Chelsea Green.
3. Meadows, Donella H., March 19–20, 1993, presentation at the conference "Towards a Sustainable Maine," Bowdoin College, Brunswick, Maine.
4. Tibbs, Hardin D. C. T., 1992, "Industrial Ecology: An Environmental Agenda for Industry," *Whole Earth Review*, Winter 1992, p. 5.
5. Renner, Michael, 1992, "Creating Sustainable Jobs in Industrial Countries," *State of the World 1992*, New York: W. W. Norton, p. 139.
6. Daly, Herman, 1990, "Toward Some Operational Principles of Sustainable Development," *Ecological Economics*, 2, p. 4.
7. Personal communication, Vermont Natural Resources Council, April 1993.
8. Tibbs, 1992, p. 5.
9. Tibbs, 1992, p. 8.
10. Tibbs, 1992, p. 4.
11. Tibbs, 1992, p. 4.
12. Schidheiny, Stephen, 1992, *Changing Course, A Global Business Perspective on Development and the Environment*, Cambridge, Mass.: MIT Press.
13. De Vaul, Diane, and Charles Bartsch, 1993, "How Utilities Can Revitalize Industry," *Issues in Science and Technology*, Spring 1993, p. 52.
14. Conservation Law Foundation et al., June 1992, *Power to Spare II: Energy Efficiency and New England's Economic Recovery, Executive Summary*, Boston, p. 1.
15. Conservation Law Foundation et al., June 1992, p. 1–2.
16. Conservation Law Foundation, 1991, *Waste Not: Garbage as an Economic Resource for the Northeast*, Boston, p. 1.
17. Conservation Law Foundation draft analysis of the potential for paper recycling in the Northern Forest region, unpublished.
18. Davis, Alan, and Susan Kinsella, May/June 1990, "Recycled Papers, Exploding the Myths," *Garbage*; Sisler, Gordon, undated, "Opportunities for Fine Paper Products," Noranda Forest Recycled Papers, p. 9.

19. Congress of the United States, Office of Technology Assessment, 1989, *Facing America's Trash: What's Next for Municipal Solid Waste?*, p. 143; *State of the World 1990*, 1990, p. 182.

20. Congress of the United States, Office of Technology Assessment, 1989, p. 142.

21. Smith, Leonard S., 1993, "Paper and Allied Products," *U.S. Industrial Outlook 1993*, p. 10–11.

22. Quigley, Jim, 1988, "Employment Impacts of Recycling," *Bicycle*, March 1988, pp. 44–47.

23. Veverka, Arthur C., 1990. "Economics Favor Increased Use of Recycled Fiber in Most Furnishes," *Pulp and Paper*, September 1990, pp. 99–100.

24. Gammie, A. P., April 23, 1992, Remarks at Groundbreaking Ceremony, East Millinocket, ME.

25. Einspahr, Dean W., 1992, "Saving Forest Resources Through Recycling," *Progress in Paper Recycling*, May 1992, p. 62.

26. North Country Regional Economic Development Council and the Adirondack North Country Association, May 1991, "A Wood Products Strategy for Northern New York, Executive Summary," p. 5.

27. Olson, Jeffrey T., and Larry Tuttle, 1990, "An Environmental Timber Solution," *Oregon Business*, April 1990, p. 35.

28. The Irland Group, July 1992, "Economic Impact of Forest Resources in Northern New England and New York: A Draft Report to the Northeastern Forest Alliance," p. 2.

29. The Irland Group, 1992, p. 2.

30. Land and Water Associates, "The Potential for Recreation Enhancements on the West Branch," submitted as Appendix B in a letter from the Conservation Law Foundation et al. to the Federal Energy Regulatory Commission on February 29, 1992, regarding the hydropower relicensing application submitted by Great Northern Nekoosa/Bowater for dams on the West Branch of the Penobscot River in Maine.

31. *Northern Forest Forum*, 1(3), Spring Equinox 1993, pp. 6, 10.

32. Hunter, Malcolm L., Jr., and Sharon Haines, 1993, "An Ecological Reserve System For the Northern Forest Lands of New England and New York: A Draft Briefing Paper for the Northern Forest Lands Council," p. 2.

33. Noss, Reed F., 1993, "Land Conservation Strategy," *Wild Earth*, a Wildlands Project Special Issue: *Plotting a North American Wilderness Recovery Strategy*, 1993, p. 15.

34. Ryan, John C., 1992, "Conserving Biological Diversity," *State of the World 1992*, New York: W. W. Norton, p. 18.

35. Hunter and Haines, 1993, p. 2.

36. Roosevelt, Theodore, 1909, *Proceedings of a Conference of Governors: In the White House, May 13–15*, Washington, D.C.: Government Printing Office, p. 10.

37. 16 U.S.C. Section 797.

38. The points that follow primarily summarize the "Platform for Reforming Federal Hydropower Policy and Restoring Environmental and Recreational Values to Rivers Developed for Hydropower" endorsed by roughly 30 conservation and river organizations from the local to national levels.

39. Shands, William E., and Robert G. Healy, 1977, *The Lands Nobody Wanted*, Conservation Foundation, p. xiii.

40. U.S. Department of Agriculture, Forest Service, 1986, *Draft Land and Resource Management Plan, Green Mountain National Forest*, p. 4.01.

41. Conservation Law Foundation et al., August 1992, "Appeal of the Record of Decision for the United States Department of Agriculture-Forest Service by the Regional Forester of the Eastern Region, Including the Land and Resource Management Plan and Final Environmental Impact Statement for the White Mountain National Forest," pp. 1–16.

42. Conservation Law Foundation et al., August 1992, p. 4, using calculations from Affidavit of V. Alaric Sample, Jr. (August 22, 1986), Appendix A-1 of CLF Appeal.

43. U.S. Department of Agriculture, Forest Service, 1986, *White Mountain National Forest Land and Water Resource Plan, Final Environmental Impact Statement*, II-101; Record of Decision: pp. 19, 20, 21, 25; FEIS: pp. II-61, 67, 77; III-12, 13; IV-9, 53; IX-B-47, 135; IX-K-73, 115.

44. U.S. Department of Agriculture, Forest Service, July 1985, "Review of Administrative Decision by the Chief of the Forest Service Related to the Forest Plans and EISs for the San Juan National Forest and the Grand Mesa, Uncompahgre, and Gunnison National Forests," p. 9.

45. Peirce, Neal R., 1993, Address to the New Hampshire Charitable Foundation, June 2, 1993.

46. Lewis, Sanford J., 1993, "Eight Principles of Genuine Partnership and Empowerment for Sustainable Industrial Development," produced by The Good Neighbor Project for Sustainable Industries (Boston), p. 9.

47. Flavin, Christopher, and John E. Young, 1993, "Shaping the Next Industrial Revolution," *State of the World 1993*, New York: W. W. Norton, p. 181.

48. Stradling, Richard, 1993, "Environmental Laws Scrutinized Study: Business Isn't Hurt," *Concord Monitor* (January 3, 1993), p. B-1.

49. Stradling, 1993, p. B-1.

50. Conservation Law Foundation et al., June 1992, p. 11.

51. Dubroff, Howard, February 1, 1993, "A Report to the Northern Forest Lands Council on Federal Taxation Issues Affecting Private Timberland Owners," p. 115.

52. 16 U.S.C. Section 502.

53. 16 U.S.C. Section 544.

10. A Sustainable Resource for a Sustainable Rural Economy

1. Harper, Stephen C., Laura L. Falk, and Edward W. Rankin, 1990, *The Northern Forest Lands Study of New England and New York*, Rutland, Vt.: U.S. Department of Agriculture, Forest Service, pp. 2–3.

2. Salwasser, Hal, Douglas W. MacCleery, and Thomas Snellgrove, 1992, "A Perspective on People, Wood, Wildlife, and Environment in U.S. Forest Stewardship," *The George Wright FORUM*, 8(4), p. 22.

3. Salwasser, MacCleery, and Snellgrove, 1992, p. 22.

4. Salwasser, MacCleery, and Snellgrove, 1992, p. 22.

5. Schallau, Con H., 1991, *State of Timber Supply—Is the Nation Positioned for the 21st Century*, Washington, D.C.: American Forest Resource Alliance, Forest Resources Technical Bulletin 9005, p. 3.

6. *Forest Resource Fact Book*, January 15, 1990, Memphis, Tenn.: National Hardwood Lumber Association, Fact Sheet 12, pp. 2, 13.

7. *Forest Statistics of the United States, 1987*, 1989, U.S. Department of Agriculture, Forest Service, Resource Bulletin PNW-168, p. 5.

8. Salwasser, MacCleery, and Snellgrove, 1992, p. 33.

9. World Commission on Environment and Development, 1987, *Our Common Future*, New York: Oxford University Press.

10. Frieswyk, Thomas S., and Anne M. Mally, 1985, *Forest Statistics for Vermont 1973 and 1983*, Radnor, Pa.: U.S. Department of Agriculture, Forest Service, Resource Bulletin NE-87, p. 38.

11. *Vermont Forest Products Industry—A Directory of Loggers and Truckers*, 1990, Waterbury, Vt.: Vermont Agency of Natural Resources, Department of Forests, Parks, and Recreation; *A Directory of Sawmills and Veneer Mills*, 1990, Waterbury, Vt.: Vermont Agency of Natural Resources, Department of Forests, Parks, and Recreation; *A Directory of Manufacturers and Craftsmen*, 1990, Waterbury, Vt.: Vermont Agency of Natural Resources, Department of Forests, Parks, and Recreation.

12. Harper, Falk, and Rankin, 1990, p. 1.

13. DeGraaf, Richard M., Mariko Yamasaki, William B. Leak, and John W. Lanier, 1992, *New England Wildlife: Management of Forested Habitats*, Radnor, Pa.: U.S. Department of Agriculture, Forest Service, General Technical Report NE-144; Gullion, Gordon W., 1984, *Managing Northern Forests for Wildlife*, St. Paul, Minn.: University of Minnesota Agricultural Experiment Station; Hornbeck, James W., and William B. Leak, 1992, *Ecology and Management of Northern Hardwood Forests in New England*, Radnor, Pa.: U.S. Department of Agriculture, Forest Service, General Technical Report NE-159; *New Perspectives on Silvicultural Management of Northern Hardwoods*, 1988, Proceedings of the 1988 Symposium on the Conflicting Consequences of Practicing Northern Hardwood Silviculture (9–10 June 1988, University of New Hampshire, Durham, N.H.), Broomall, Pa.: U.S. Department of Agriculture, Forest Service, General Technical Report NE-124; Seymour, Robert S., and Malcolm L. Hunter, Jr., 1992, *New Forestry in Eastern Spruce-Fir Forests: Principles and Applications to Maine*, Orono, Maine: University of Maine, Maine Agricultural Experimental Station, Misc. Publication 716.

14. Newton, Carlton M., et al., 1990, Executive Summary of *Impact Assessment of Timber Harvesting Activity in Vermont*, Burlington, Vt.: University of Vermont, School of Natural Resources.

15. Newton et al., 1990, p. 7.

16. United States Forest Service, 1979, *A Guide to Common Insects and Diseases of Forest Trees in the Northeastern United States*, Broomall, Pa.: U.S. Department of Agriculture, Forest Service, Forest Insect and Disease Management NA-FR-4.

17. New England Society of American Foresters position summary on the U.S. Forest Service Northern Forest Land Study Area, 1989, Montpelier, Vt.: New England Society of American Foresters.

11. Northern Appalachian Wilderness

1. Merchant, Carolyn, 1989, *Ecological Revolutions: Nature, Gender, and Science in New England*, Chapel Hill, N.C.: University of North Carolina Press, p. 101.

2. Turner, Frederick, 1986, *Beyond Geography: The Western Spirit Against the Wilderness*, New Brunswick, N.J.: Rutgers University Press, p. 216.

3. Perlin, John, 1989, *A Forest Journey: The Role of Wood in the Development of Civilization*, New York: W. W. Norton, p. 314.
4. Smith, David C., 1972, *A History of Lumbering in Maine, 1861–1960*, Orono, Maine: University of Maine Studies, p. 174.
5. Brown, William R., 1958, *Our Forest Legacy*, Concord, N.H.: New Hampshire Historical Society, p. 6.
6. Brown, 1958, pp. 21–22.
7. Smith, 1972, p. 4.
8. Roszak, Theodore, 1978, *Person/Planet: The Creative Disintegration of Industrial Society*, Garden City, New York: Anchor/Doubleday, p. 261.
9. Brown, 1958, p. 118.
10. Norberg-Hodge, Helena, 1991, *Ancient Futures: Learning from Ladakh*, San Francisco: Sierra Club Books, p. 155.
11. Smith, 1972, p. 18.
12. Holbrook, Stewart, 1980, *Holy Old Mackinaw: A Natural History of the American Lumberjack*, Sausalito, Calif.: Comstock Edition, p. 51.
13. Lansky, Mitch, 1992, *Beyond the Beauty Strip: Saving What's Left of Our Forests*, Gardiner, Maine: Tilbury House, pp. 77–78.
14. Lansky, 1992, p. 44.
15. Lansky, 1992, p. 87.
16. Lansky, 1992, p. 47; also Champion International, New York Region, Timberlands Staff, April 1989, "Changing The Landowners' Economic Conditions: A New Hampshire/Vermont Case Study of Champion International Corporation," paper for Northern Forest Lands Study, p. 8.
17. Lansky, 1992, p. 80.
18. Lansky, 1992, pp. 64–65.
19. Smith, 1972, p. 109.
20. Smith, 1972, p. 21.
21. Champion International, New York Region, Timberlands Staff, 1989, p. 12; Blackmer, Steve, July 1985, "The Export of Sawlogs from New York and New England to Canada," paper prepared for SELL-USA, p. 7.
22. Holbrook, 1980, p. 145.
23. Smith, 1972, pp. 88, 292.
24. Lansky, 1992, pp. 204–210.
25. Lansky, 1992, pp. 370–372.
26. In autumn 1987 this long-term shift in industry strategy that would result in the sale of millions of acres of industrial forest land was predicted by respected industry analyst Perry Hagenstein, 1987, "A Challenge for New England: Changes in Large Forest Land Holdings," Boston: The Fund for New England.
27. Meadows, Donella, 1991, *The Global Citizen*, Washington, D.C.: Island Press, p. 33.
28. Humbach, John A., "Law and a New Land Ethic," in *Proceedings of the Second Annual Growth Management Forum: Sustaining the Economic and Environmental Future of New England*, Medford, Mass.: Tufts University, Lincoln Filene Center, Environmental Affairs Program, pp. 6–25; Natural Resources Council of Maine, September 19, 1991, "A North Woods Conservation Area: Preserving Maine's Traditional Forest Uses," Augusta, Maine.
29. See David Carle, 1992, "Lake Umbagog Easement—Have Public Expectations Been Met?" and "Easements: Advantages and Disadvantages," *The Northern Forest Forum*, Winter Solstice, pp. 6–7.
30. Some have gone so far as to suggest that there are alternatives to toilet paper, a product that was only invented at the end of the last century. See Streeter, Robert, "Wiping for Wilderness," *Glacial Erratic*, Summer 1990, p. 8.
31. See Merchant, 1989, pp. 149–156, for an account of the subsistence agricultural economy of Colonial New England.

32. Sayen, Jamie, Summer 1992, "Civilian Conservation Corps," *Wild Earth*, Summer 1992, p. 50.

33. Crawford, Hewlette S., and Robert M. Frank, "Wildlife Habitat Responses to Silvicultural Practices in Spruce-Fir Forests," *Transactions of the 52nd North American Wildlife and Natural Resource Conference*, p. 99.

34. Maser, Chris, 1988, *The Redesigned Forest*, San Pedro, Calif.: R & E Miles, pp. 68–69.

35. Thoreau, Henry David, 1864, reprinted 1966, *The Maine Woods*, New York: Thomas Y. Crowell Company, pp. 198–199.

36. Merchant, 1989, p. 226.

37. Trombulak, Steve, 1992, "Amphibians of the Northern Forests," *The Northern Forest Forum*, Autumn Equinox, pp. 4–5.

38. Noss, 1993, p. 11.

39. Noss, Reed, 1993, "The Wildlands Project: Land Conservation Strategy," *Wild Earth, Special Wildlands Project Issue*, Winter, p. 15.

40. Sayen, Jamie, May 1, 1987, "The Appalachian Mountains: Vision and Wilderness," *Earth First!*, May 1, 1987, pp. 26–30.

41. Hunter, Malcolm, G. L. Jacobson, and T. Webb, 1988. "Paleoecology and the Coarse-Filter Approach to Maintaining Biological Diversity," *Conservation Biology*, 2:375–385.

42. Davis, George, 1988, *2020 Vision—Fulfilling the Promise of the Adirondack Park, Volume I: Biological Diversity: Saving All the Pieces*, October, The Adirondack Council (see especially p. 5).

43. Lynn, Evadna S., September 21, 1992, "A Wall Street Perspective on National/International Influences on Land Ownership and Conversion in Northern Forest Lands," presentation to Northern Forest Lands Council, p. 4.

44. The account of the September 21, 1992, forum of the Land Conversion Subcommittee appeared in a somewhat different form in Sayen, Jamie, 1992, "Bleak Paper Industry Future: A Chance to Buy Millions of Acres of the Northern Forests," *The Northern Forest Forum*, Winter Solstice, p. 5.

45. Harper, Stephen C., Laura L. Falk, and Edward W. Rankin, 1990, *The Northern Forest Lands Study of New England and New York*, Rutland, VT: U.S. Department of Agriculture, Forest Service, p. 49.

46. Krassner, Lowell, 1992, "Vermont Town Buys 1639 Acres," *The Northern Forest Forum*, Winter Solstice, p. 5.

Index